Quantities, Units and Symbols in Physical Chemistry

4th Edition, Abridged Version

Quantities, Units and Symbols in Physical Chemistry
4th Edition, Abridged Version

INTERNATIONAL UNION OF PURE AND APPLIED CHEMISTRY
Physical and Biophysical Chemistry Division

Prepared for publication by

Christopher M. A. Brett
University of Coimbra, Portugal
Email: cbrett@ci.uc.pt

Jeremy G. Frey
University of Southampton, UK
Email: j.g.frey@soton.ac.uk

Robert Hinde
The University of Tennessee, USA
Email: rhinde@utk.edu

Yutaka Kuroda
Tokyo University of Agriculture and Technology, Japan
Email: ykuroda@cc.tuat.ac.jp

Roberto Marquardt
Université de Strasbourg, France
Email: roberto.marquardt@unistra.fr

Franco Pavese
Instituto Nazionale di Ricerca Metrologica, Italy
Email: F.Pavese@gmail.com

Martin Quack
ETH Zurich, Switzerland
Email: quack@ir.phys.chem.ethz.ch

Jürgen Stohner
Zurich University of Applied Sciences, Switzerland
Email: sthj@zhaw.ch

and

Anders J Thor
Deceased
Stockholm, Sweden

ROYAL SOCIETY OF CHEMISTRY

IUPAC

INTERNATIONAL UNION OF PURE AND APPLIED CHEMISTRY

Print ISBN: 978-1-83916-150-6
PDF ISBN: 978-1-83916-318-0

A catalogue record for this book is available from the British Library

The Royal Society of Chemistry is a charity, registered in England and Wales, Number 207890, and a company incorporated in England by Royal Charter (Registered No. RC000524), registered office: Burlington House, Piccadilly, London W1J 0BA, UK, Telephone: +44 (0) 20 7437 8656.

Visit our website at www.rsc.org/books

Printed in the United Kingdom by CPI Group (UK) Ltd, Croydon, CR0 4YY, UK

CONTENTS

CONTENTS v

PREFACE vii

HISTORICAL INTRODUCTION viii

1 PHYSICAL QUANTITIES AND UNITS 1
- 1.1 Physical quantities and quantity calculus . . . 1
- 1.2 Base quantities and derived quantities . . . 2
- 1.3 Symbols for physical quantities and units . . . 2
  - 1.3.1 General rules for symbols for quantities . . . 3
  - 1.3.2 General rules for symbols for units . . . 3
- 1.4 Use of the words "extensive", "intensive", "specific", and "molar" . . . 4
- 1.5 Products and quotients of physical quantities and units . . . 5
- 1.6 The use of italic and roman fonts for symbols in scientific publications . . . 5
- 1.7 Printing of numbers and mathematical symbols . . . 7
- 1.8 Symbols, operators, and functions . . . 8

2 DEFINITIONS AND SYMBOLS FOR UNITS 11
- 2.1 The International System of Units (SI) . . . 11
- 2.2 Names and symbols for the SI base units . . . 11
- 2.3 Coherent units and checking dimensions . . . 12
- 2.4 Physical constants used as atomic units . . . 12
- 2.5 SI prefixes and prefixes for binary multiples . . . 14

3 CONVERSION OF UNITS 15
- 3.1 The use of quantity calculus . . . 15
- 3.2 Conversion tables for units . . . 16
- 3.3 Transformation of equations of electromagnetic theory between the ISQ (SI) and Gaussian forms . . . 20

4 TABLES OF PHYSICAL QUANTITIES 21
- 4.1 Space and time . . . 22
- 4.2 Classical mechanics . . . 23
- 4.3 Electricity and magnetism . . . 24
- 4.4 Quantum mechanics and quantum chemistry . . . 25
  - 4.4.1 Symbols commonly used in quantum chemistry . . . 26
- 4.5 Atoms and molecules . . . 27
- 4.6 Spectroscopy . . . 29
  - 4.6.1 Optical spectroscopy . . . 29
  - 4.6.2 Electron paramagnetic resonance (EPR), electron spin resonance (ESR), and nuclear magnetic resonance (NMR) . . . 31
  - 4.6.3 Symbols for angular momentum operators and quantum numbers . . . 33
  - 4.6.4 Symbols for symmetry operators and labels for symmetry species . . . 33
  - 4.6.5 Other symbols and conventions in optical spectroscopy . . . 34
- 4.7 Electromagnetic radiation . . . 36
  - 4.7.1 Quantities and symbols concerned with the measurement of absorption . . . 39
  - 4.7.2 Practical equations linking A_{fi}, G_{fi} and M_{fi} as well as f_{fi} . . . 40
- 4.8 Solid state . . . 41
  - 4.8.1 Symbols for planes and directions in crystals . . . 42
- 4.9 Statistical thermodynamics . . . 43

4.10 General chemistry . . . 44
4.10.1 Other symbols and conventions in chemistry . . . 45
4.11 Chemical thermodynamics . . . 49
4.11.1 Other symbols and conventions in chemical thermodynamics . . . 51
4.12 Chemical kinetics and photochemistry . . . 54
4.12.1 Other symbols, terms, and conventions used in chemical kinetics . . . 57
4.13 Electrochemistry . . . 59
4.13.1 Sign and notation conventions in electrochemistry . . . 61
4.14 Colloid and surface chemistry . . . 63
4.15 Transport properties . . . 64
4.15.1 Transport characteristic numbers: Quantities of dimension one . . . 65

5 UNCERTAINTY **67**
5.1 Reporting uncertainty for a single measured quantity . . . 67

6 REFERENCES **69**
6.1 Primary sources . . . 69
6.2 Secondary sources . . . 71

7 GREEK ALPHABET **76**

8 INDEX OF SYMBOLS **77**

9 SUBJECT INDEX **87**

PRESSURE CONVERSION FACTORS **108**

NUMERICAL ENERGY CONVERSION FACTORS **109**

FUNDAMENTAL PHYSICAL CONSTANTS **inside front cover**

IUPAC PERIODIC TABLE OF THE ELEMENTS **inside back cover**

PREFACE

This manual is an abridged version of "Quantities, Units and Symbols in Physical Chemistry", the "Green Book" of the International Union of Pure and Applied Chemistry(IUPAC). It is based on the 4th edition of the Green Book, which is to be published. The 3rd edition of the Green Book was published by the Royal Society of Chemistry, Cambridge UK, in 2007, and has since then been reprinted in revised form. The manual was prepared with the aim to provide students and workers with good examples of the use of quantities, symbols and units widely employed in physics and physical chemistry. As its original main work, it also has the purpose to improve the exchange of scientific information among readers in different disciplines and across different nations. The importance of a good practice in the use of terminology, quantities, units and their symbols must be emphasized. In times when information gets more and more tweeted, "liked" or "disliked", rather than conveyed in clear sentences and formulae, it is of the utmost importance to ensure the correct dissemination of accurate scientific information.

Physical chemistry is transdisciplinary by its very nature and thus this manual is addressed to students from all disciplines. It offers them a rapid access to quantities, their units and symbols with selected examples of the good practice of their usage. The good use of quantities and units is important in the medical domain in particular. Recently, while undergoing a routine examination with the ophthalmologist, I was told that my eye pressure had a normal level. Being asked what this meant, the doctor said: "15". At that point my eyes began to flash and the doctor, knowing my profession, added "millimeter mercury". This episode is perhaps just fun for some, others might not yet have seen the point. The misuse of units, in particular the neglect to even mention them, can decide about life and death. In another episode, reported 2007 by Tomaszewski in the Journal of Medical Toxicology, volume 3, page 87, a patient receiving treatment in a hospital was supposedly in need of glucose, as the apparatus to determine the glucose level was apparently indicating a level of 42 mg/dl, when (cit.) "inspection of the glucose meter revealed an actual reading of 42 mmol/L, in other words, an extraordinarily high glucose of 758 mg/dl, more consistent with hyperosmolar coma than hypoglycemia."

Other examples arising from a misuse of inconsistent units, such as that of the Mars orbiter crash in 1999, or the Laufenburg bridge mishap in 2003, and many others can be mentioned (see, for instance, Lewis, PRIMUS, 25, 181 (2015)). The key message told by all these examples is the same: the value of a physical quantity is composed of a numerical value and a unit; it is actually the product of these two factors and **these two factors must always be given in the report of the value of a physical quantity**. The use of units from the International System of Units (SI) can be helpful and is recommended. Other units can be used, if the rule is observed that units must always be well defined and added as co-factors to the numerical value of a physical quantity. If units other than SI units are used, they should be clearly defined in relation to the SI units.

The interpretation of physical quantities as being a product of the numerical value and the unit leads to *quantity calculus*, a mathematical method to treat without failure physical quantities that goes back to James Clark Maxwell in "A Treatise on Electricity and Magnetism", Oxford, Clarendon Press, vol 1, page 1, 1873. It is another purpose of this manual to recall and reinforce the use of quantity calculus in scientific disciplines.

During the final preparation of this manuscript we learned that Prof. Kozo Kuchitsu deceased. Prof. Kuchitsu was a founding author of the Green Book from its first edition and we dedicate this book to his memory.

Prof. Roberto Marquardt
Chairman of the task group working in the IUPAC project 2007-032-1-100
Strasbourg, May 2023

HISTORICAL INTRODUCTION

The *Manual of Symbols and Terminology for Physicochemical Quantities and Units* [1.a], to which the 'Green Book' is a successor, was first prepared for publication on behalf of the Physical Chemistry Division of IUPAC by M. L. McGlashan in 1969, when he was chairman of the Commission on Physicochemical Symbols, Terminology and Units (I.1). He made a substantial contribution towards the objective which he described in the preface to that first edition as being "to secure clarity and precision, and wider agreement in the use of symbols, by chemists in different countries, among physicists, chemists and engineers, and by editors of scientific journals". The second edition of that manual prepared for publication by M. A. Paul in 1973 [1.b], and the third edition prepared by D. H. Whiffen in 1976 [1.c], were revisions to take account of various developments in the Système International d'Unités (International System of Units, international abbreviation SI), and other developments in terminology.

The first edition of *Quantities, Units and Symbols in Physical Chemistry* published in 1988 [2.a] became widely known as the 'Green Book' of IUPAC and was a substantially revised and extended version of the earlier manuals. The decision to embark on this project originally proposed by N. Kallay was taken at the IUPAC General Assembly at Leuven in 1981, when D. R. Lide was chairman of the Commission. The working party was established at the 1983 meeting in Lyngby, when K. Kuchitsu was chairman, and the project has received strong support throughout from all present and past members of the Commission I.1 and other Physical Chemistry Commissions, particularly D. R. Lide, D. H. Whiffen, and N. Sheppard.

The extensions included some of the material previously published in appendices [1.d–1.k]; all the newer resolutions and recommendations on units by the Conférence Générale des Poids et Mesures (CGPM); and the recommendations of the International Union of Pure and Applied Physics (IUPAP) of 1978 and of Technical Committee 12 of the International Organization for Standardization, *Quantities, units, symbols, conversion factors* (ISO/TC 12). The tables of physical quantities (Chapter 2) were extended to include defining equations and SI units for each quantity. The style was also slightly changed from being a book of rules towards a manual of advice and assistance for the day-to-day use of practicing scientists. Examples of this are the inclusion of extensive notes and explanatory text inserts in Chapter 2, the introduction to quantity calculus, and the tables of conversion factors between SI and non-SI units and equations in Chapter 7.

The second edition (1993) [2.b] was a revised and extended version of the previous edition. The revisions were based on the recent resolutions of the CGPM (see history of [3]); the new recommendations by IUPAP [4]; the new international standard ISO 31 [5,6]; some recommendations published by other IUPAC commissions; and numerous comments we have received from chemists throughout the world. The revisions in the second edition were mainly carried out by I. Mills and T. Cvitaš with substantial input from R. Alberty, K. Kuchitsu, M. Quack as well as from other members of the IUPAC Commission on Physicochemical Symbols, Terminology and Units.

The manual has found wide acceptance in the chemical community, and various editions have been translated into Russian [2.d], Hungarian [2.e], Japanese [2.f], German [2.g], Romanian [2.h], Spanish [2.i], Catalan [2.j], French [2.k], and Portuguese [2.l]. Large parts of it have been reproduced in the 71st and subsequent editions of the *Handbook of Chemistry and Physics* published by CRC Press.

The work on revisions of the second edition started immediately after its publication and between 1995 and 1997 it was discussed to change the title to "Physical-Chemical Quantities, Units and Symbols" and to apply rather complete revisions in various parts. It was emphasized that the book covers as much the field generally called "physical chemistry" as the field called "chemical physics". Indeed we consider the larger interdisciplinary field where the boundary between physics and chemistry has largely disappeared [10]. At the same time it was decided to produce the whole book as a text file in computer readable form to allow for future access directly by computer, some time after the printed version would be available. Support for this decision came from the IUPAC secretariat in the Research Triangle Park, NC (USA) (John W. Jost). The practical work on the revisions was carried out at the ETH Zürich, while the major input on this edition came from the group of editors listed now in full on the cover. It fits with the new structure of IUPAC that these are defined as project members and not only through membership in the commission. The basic structure of this edition was finally established at a working meeting of the project members in Engelberg, Switzerland in March 1999, while further revisions were discussed at the Berlin meeting (August 1999) and thereafter. In 2001 it was decided finally to use the old title. In this third edition [2.c] the whole text and all tables have been revised, many chapters substantially. This work was carried out mainly at ETH Zürich, where Jürgen Stohner coordinated the various contributions and corrections from the project group members and prepared the print-ready electronic document. Larger changes compared to previous editions concern a complete and substantial update of recently available improved constants, sections on uncertainty in physical quantities, dimensionless quantities, mathematical symbols and numerous other sections, as well as a much improved index.

Between 2007 and 2020 the fourth edition has been prepared, which now includes also the changes to the SI, which became effective in May 2019. While everything has been updated accordingly in the 4th edition, including also new values for the physical constants and other related quantities, the basic structure of the book has been retained. While the publication of the 4th edition is in progress, we present here as an initial publication the 'Abridged Version'.

At the end of this historical survey we might refer also to what might be called the tradition of this manual. It is not the aim to present a list of recommendations in form of commandments. Rather, we have always followed the principle that this manual should help the user in what may be called "good practice of scientific language". If there are several well established uses or conventions, these have been mentioned, giving preference to one, when this is useful, but making allowance for variety, if such variety is not harmful to clarity. In a few cases possible improvements to conventions or language are mentioned with appropriate reference, even if uncommon, but without specific recommendation. In those cases where certain common uses are deprecated, there are very strong reasons for this and the reader should follow the corresponding advice.

Zürich, 2023 Martin Quack

The membership of the Commission during the period 1963 to 2023, during which the successive editions of this manual were prepared, was as follows (selected list, not complete):

Titular members

Chairman: 1963–1967 G. Waddington (USA); 1967–1971 M.L. McGlashan (UK); 1971–1973 M.A. Paul (USA); 1973–1977 D.H. Whiffen (UK); 1977–1981 D.R. Lide Jr (USA); 1981–1985 K. Kuchitsu (Japan); 1985–1989 I.M. Mills (UK); 1989–1993 T. Cvitaš (Croatia); 1993–1999 H.L. Strauss (USA); 2000–2007, 2018–2019 J.G. Frey (UK); 2008–2011 R. Marquardt (France); 2012–2017 J. Stohner (Switzerland); 2020 R. Weir (Canada); 2021 A.J. McQuillan (New Zealand); 2022–2023 Y. Kuroda (Japan).

Secretary: 1963–1967 H. Brusset (France); 1967–1971 M.A. Paul (USA); 1971–1975 M. Fayard (France); 1975–1979 K.G. Weil (Germany); 1979–1983 I. Ansara (France); 1983–1985 N. Kallay (Croatia); 1985–1987 K.H. Homann (Germany); 1987–1989 T. Cvitaš (Croatia); 1989–1991 I.M. Mills (UK); 1991–1997, 2001–2005 M. Quack (Switzerland); 1997–2001 B. Holmström (Sweden); 2008–2011 J. Stohner (Switzerland); 2012–2017 R. Hinde (USA); 2018–2019 R. Rocha Filho (Brazil); 2020–2023 R. Marquardt (France).

Other titular members

1975–1983 I. Ansara (France); 1965–1969 K.V. Astachov (Russia); 1963–1971 R.G. Bates (USA); 1963–1967 H. Brusset (France); 1985–1997 T. Cvitaš (Croatia); 1963 F. Daniels (USA); 1979–1981 D.H.W. den Boer (Netherlands); 1981–1989 E.T. Denisov (Russia); 1967–1975 M. Fayard (France); 1997–2007, 2018–2019 J. Frey (UK); 1963–1965 J.I. Gerassimov (Russia); 2014–2015 A. Goodwin (USA); 2012–2017 R. Hinde (USA); 1991–2001 B. Holmström (Sweden); 1979–1987 K.H. Homann (Germany); 1963–1971 W. Jaenicke (Germany); 1967–1971 F. Jellinek (Netherlands); 1977–1985 N. Kallay (Croatia); 2020–2023 J. Kaiser (UK); 1973–1981 V. Kellö (Czechoslovakia); 1989–1997 I.V. Khudyakov (Russia); 1985–1987 W.H. Kirchhoff (USA); 1971–1979 J. Koefoed (Denmark); 1979–1987 K. Kuchitsu (Japan); 2012–2015, 2020–2023 Y. Kuroda (Japan); 1971–1981 D.R. Lide Jr (USA); 1997–2001, 2006–2011, 2020–2023 R. Marquardt (France); 1963–1971 M.L. McGlashan (UK); 2012–2015, 2018–2023 A.J. McQuillan (New Zealand); 1983–1991 I.M. Mills (UK); 1963–1967 M. Milone (Italy); 1967–1973 M.A. Paul (USA); 1991–1999, 2006–2009 F. Pavese (Italy); 1963–1967 K.J. Pedersen (Denmark); 1967–1975 A. Perez-Masiá (Spain); 1987–1997, 2001–2005 M. Quack (Switzerland); 1971–1979 A. Schuyff (Netherlands); 1967–1970 L.G. Sillén (Sweden); 1997–2001, 2008–2017 J. Stohner (Switzerland); 1989–1999, 2002–2005 H.L. Strauss (USA); 1995–2001 M. Takami (Japan); 1987–1991 M. Tasumi (Japan); 1963–1967 G. Waddington (USA); 1981–1985 D.D. Wagman (USA); 1971–1979 K.G. Weil (Germany); 2020 R. Weir (Canada); 1971–1977 D.H. Whiffen (UK); 1963–1967 E.H. Wiebenga (Netherlands).

Associate members

1983–1991 R.A. Alberty (USA); 1983–1987 I. Ansara (France); 2020–2023 S. Chalk (USA); 1979–1991 E.R. Cohen (USA); 2020–2023 G. Deng (China); 1979–1981 E.T. Denisov (Russia); 1987–1991 G.H. Findenegg (Germany); 1987–1991 K.H. Homann (Germany); 1971–1973 W. Jaenicke (Germany); 1985–1989 N. Kallay (Croatia); 1987–1989 and 1998–1999 I.V. Khudyakov (Russia); 1979–1980 J. Koefoed (Denmark); 1987–1991 K. Kuchitsu (Japan); 1981–1983 D.R. Lide Jr (USA); 1971–1979 M.L. McGlashan (UK); 1991–1993 I.M. Mills (UK); 1973–1981 M.A. Paul (USA); 1999–2005, 2016–2019 F. Pavese (Italy); 1975–1983 A. Perez-Masiá (Spain); 1997–1999 M. Quack (Switzerland); 1979–1987 A. Schuyff (Netherlands); 1963–1971 S. Seki (Japan); 2000–2001 H.L. Strauss (USA); 1991–1995 M. Tasumi (Japan); 1969–1977 J. Terrien (France); 1994–2001 A J Thor (Sweden); 1975–1979 L. Villena (Spain); 1967–1969 G. Waddington (USA); 1979–1983 K.G. Weil (Germany); 1977–1985 D.H. Whiffen (UK).

National representatives

Numerous national representatives have served on the Commission over many years. We do not provide this long list here.

At the time of the publication of the abridged version (2023) of the 4th edition, the Commission had the following composition:

Y. Kuroda (Chairman), R. Marquardt (Secretary), J. McQuillan (TM), J. Kaiser (TM), S. Chalk (AM), G. Deng (AM), M. Quack (NR Switzerland), P. Metrangolo (ex officio). We acknowledge also the contributions from numerous persons who were not involved in any of the Commission's functions. There have been also considerable efforts in generating new translations of the 3rd edition of the 'Green Book' which are acknowledged [2.k, 2.l, 2.m].

1 PHYSICAL QUANTITIES AND UNITS

1.1 PHYSICAL QUANTITIES AND QUANTITY CALCULUS

The value of a *physical quantity* Q can be expressed as the product of a *numerical value* $\{Q\}$ and a *unit* $[Q]$

$$Q = \{Q\}\ [Q] \tag{1}$$

Neither the name of the physical quantity, nor the symbol used to denote it, implies a particular choice of unit.

Physical quantities, numerical values, and units may all be manipulated by the ordinary rules of algebra. Thus we may write, for example, for the wavelength λ of one of the yellow sodium lines

$$\lambda = 5.896 \times 10^{-7}\ \mathrm{m} = 589.6\ \mathrm{nm} \tag{2}$$

where m is the symbol for the unit of length called the metre (see Section 2.2, p. 11), nm is the symbol for the nanometre, and the units metre and nanometre are related by

$$1\ \mathrm{nm} = 10^{-9}\ \mathrm{m} \qquad \text{or} \qquad \mathrm{nm} = 10^{-9}\ \mathrm{m} \tag{3}$$

The equivalence of the two expressions for λ in Equation (2) follows at once when we treat the units by the rules of algebra and recognize the identity of 1 nm and 10^{-9} m in Equation (3). The wavelength may equally well be expressed in the form

$$\lambda/\mathrm{m} = 5.896 \times 10^{-7} \qquad \text{or} \qquad \lambda/\mathrm{nm} = 589.6 \tag{4}$$

It can be useful to work with variables that are defined by dividing the quantity by a particular unit. For instance, in tabulating the numerical values of physical quantities or labeling the axes of graphs, it is particularly convenient to use the quotient of a physical quantity and a unit in such a form that the values to be tabulated are numerical values, as in Equations (4).

Example

$$\ln(p/\mathrm{MPa}) = a + b/T = a + b'(10^3\ \mathrm{K}/T) \tag{5}$$

T/K	10^3 K/T	p/MPa	ln(p/MPa)
216.55	4.6179	0.5180	−0.6578
273.15	3.6610	3.4853	1.2486
304.19	3.2874	7.3815	1.9990

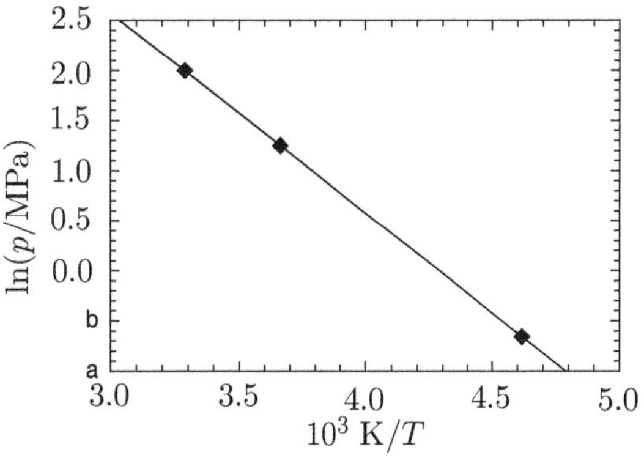

Algebraically equivalent forms may be used in place of 10^3 K/T, such as kK/T or 10^3 $(T/\mathrm{K})^{-1}$. Equations between numerical values depend on the choice of units, whereas equations between quantities have the advantage of being independent of this choice. The method described here for handling physical quantities and their units is known as *quantity calculus* (see ISO [5.a] and [11–14]). It is recommended for use throughout science and technology. The use of quantity calculus does not imply any particular choice of units (see Section 3.1, p. 15 for quantity calculus).

1.2 BASE QUANTITIES AND DERIVED QUANTITIES

By convention physical quantities are organized in a dimensional system built upon seven *base quantities*, each of which is regarded as having its own dimension. These base quantities in the International System of Quantities (ISQ) on which the International System of units (SI) is based, and the principal symbols used to denote them and their dimensions are as follows:

Base quantity	*Symbol for quantity*	*Symbol for dimension*
length	l	L
mass	m	M
time	t	T
electric current	I	I
thermodynamic temperature	T	Θ
amount of substance	n	N
luminous intensity	I_{v}	J

All other quantities are called *derived quantities* and are regarded as having dimensions derived algebraically from the seven base quantities by multiplication and division.

Example dimension of energy is equal to dimension of $\mathsf{M\ L^2\ T^{-2}}$
This can be written with the symbol dim for dimension (see footnote [1], below)
$\mathrm{dim}(E) = \mathrm{dim}(m \cdot l^2 \cdot t^{-2}) = \mathsf{M\ L^2\ T^{-2}}$

The quantity *amount of substance* is of special importance to chemists. Amount of substance is proportional to the number of specified elementary entities of the substance considered. The proportionality factor is the same for all substances; its reciprocal is the *Avogadro constant* (see Section 4.10, p. 44). The SI unit of amount of substance is the mole. "Amount of substance" is also called "chemical amount", and may be abbreviated to the single word "amount", particularly in such phrases as "amount concentration" (see footnote [2]), and "amount of N_2". In the same sense one might for instance say "amount concentration of N_2." A possible name for international usage has been suggested: "enplethy" [10] (from Greek, similar to enthalpy and entropy).

In the ISQ, electric current is chosen as base quantity and ampere is the SI base unit. In atomic and molecular physics, the so-called *atomic units* are useful (see Section 2.4, p. 12).

Ordinal quantities and nominal properties are outside the scope of the Green Book [9].

1.3 SYMBOLS FOR PHYSICAL QUANTITIES AND UNITS [5.a]

A clear distinction should be drawn between the names and symbols for physical quantities, and the names and symbols for units. Names and symbols for many quantities are given in Chapter 4, p. 21; the symbols given there are *recommendations*. If other symbols are used they should be clearly defined. Names and symbols for units are given in Chapter 2, p. 11; the symbols for units listed there are quoted from the Bureau International des Poids et Mesures (BIPM) and are *mandatory*.

[1] The symbol $[Q]$ was formerly used for *dimension* of Q, but this symbol is used and preferred for *unit* of Q, see Eq. (1) on page 1.

[2] The Chemistry and Human Health Division of IUPAC recommended that "amount-of-substance concentration" be abbreviated "substance concentration" [15].

1.3.1 General rules for symbols for quantities

The symbol for a physical quantity should be a capital or lower case single letter (see footnote [1]) of the Latin or Greek alphabet (see Section 1.6, p. 5). The letter should be printed in italic (sloping) type. When necessary the symbol may be modified by subscripts and superscripts of specified meaning. Subscripts and superscripts that are themselves symbols for physical quantities or for numbers should be printed in italic type; other subscripts and superscripts should be printed in roman (upright) type.

Examples	C_p	for heat capacity at constant pressure
	p_i	for partial pressure of the ith substance
but	C_B	for heat capacity of substance B
	$\mu_\mathrm{B}{}^{\alpha}$	for chemical potential of substance B in phase α
	$\Delta_\mathrm{r}H^\ominus$	for standard reaction enthalpy

The meaning of symbols for physical quantities may be further qualified by the use of one or more subscripts, or by information contained in parentheses.

Examples $\Delta_\mathrm{f}S^\ominus(\mathrm{HgCl_2}, \mathrm{cr}, 25\ °\mathrm{C}) = -154.3\ \mathrm{J\ K^{-1}\ mol^{-1}}$

$\mu_i = (\partial G/\partial n_i)_{T,p,\ldots,n_j,\ldots;\, j\neq i}$ or $\mu_i = (\partial G/\partial n_i)_{T,p,n_{j\neq i}}$

Vectors and matrices may be printed in bold-face italic type, e.g. $\boldsymbol{A}, \boldsymbol{a}$. Tensors may be printed in bold-face italic sans serif type, e.g. $\boldsymbol{\mathsf{S}}$, $\boldsymbol{\mathsf{T}}$. Vectors may alternatively be characterized by an arrow, $\vec{A}$, $\vec{a}$ and second-rank tensors by a double arrow, $\vec{\vec{S}}, \vec{\vec{T}}$.

1.3.2 General rules for symbols for units

Symbols for units should be printed in roman (upright) type. They should remain unaltered in the plural, and should not be followed by a full stop except at the end of a sentence.

Example $r = 10$ cm, not cm. or cms.

When a quantity symbol is used as a symbol for a unit, it is written following the general rules for symbols for quantities (see Section 1.3.1, p. 3).

Example $r = 1\,a_0$, where a_0 is the bohr, the atomic unit of length.

Symbols for units shall be printed in lower case letters, unless they are derived from a personal name when they shall begin with a capital letter. An exception is the symbol for the litre which may be either L or l, i.e. either capital or lower case (see footnote [2]).

Examples m (metre), s (second), but J (joule), Hz (hertz)

Decimal multiples and submultiples of units may be indicated by the use of prefixes as defined in Section 2.5, p. 14.

Examples nm (nanometre), MHz (megahertz), kV (kilovolt)

---

[1] An exception is made for certain characteristic numbers or "dimensionless quantities" used in the study of transport processes for which the internationally agreed symbols consist of two letters (see Section 4.15.1, p. 65).

Example Reynolds number, *Re* (see Section 4.14.1, p. 65); another example for a symbol with two letters is pH (see Section 4.13, p. 59).

When such symbols appear as factors in a product, they should be separated from other symbols by a space, multiplication sign, or parentheses.

[2] However, only the lower case l is used by ISO and the International Electrotechnical Commission (IEC).

1.4 USE OF THE WORDS "EXTENSIVE", "INTENSIVE", "SPECIFIC", AND "MOLAR"

A quantity that is additive for independent, noninteracting subsystems is called *extensive*; examples are mass m, volume V, Gibbs energy G. A quantity that is independent of the extent of the system is called *intensive*; examples are temperature T, pressure p, chemical potential (partial molar Gibbs energy) μ.

The adjective *specific* before the name of an extensive quantity is used to mean *divided by mass*. When the symbol for the extensive quantity is a capital letter, the symbol used for the specific quantity is often the corresponding lower case letter.

Examples volume, V, and specific volume, $v = V/m = 1/\rho$ (where ρ is mass density);
heat capacity at constant pressure, C_p, and
specific heat capacity at constant pressure, $c_p = C_p/m$

ISO [5.a] and the Chemistry and Human Health Division of IUPAC recommend systematic naming of physical quantities derived by division with mass, volume, area, and length by using the attributes massic or specific, volumic, areic, and lineic, respectively. Thus, for example, the specific volume could be called massic volume and the surface charge density would be areic charge. In addition the Chemistry and Human Health Division of IUPAC recommends the use of the attribute entitic for quantities derived by division with the number of entities [15].

The adjective *molar* [1] before the name of an extensive quantity generally means *divided by amount of substance*. The subscript m on the symbol for the extensive quantity denotes the corresponding molar quantity.

Examples volume, V molar volume, $V_\mathrm{m} = V/n$ (Section 4.10, p. 44)
enthalpy, H molar enthalpy, $H_\mathrm{m} = H/n$

It is sometimes convenient to divide all extensive quantities by amount of substance, so that all quantities become intensive; the subscript m may then be omitted if this convention is stated and there is no risk of ambiguity. (See also the symbols recommended for partial molar quantities in Section 4.11, p. 49, and in Section 4.11.1 (iii), p. 51.)

Examples molar absorption coefficient, $\varepsilon = a/c$ (see Section 4.7, p. 38)
molar conductivity, $\Lambda = \kappa/c$ (see Section 4.13, p. 60)

Sometimes the adjective *molar* has a different meaning, namely *divided by amount-of-substance concentration*. The adjective *molar* is also used instead of the unit mol dm^{-3} for an amount concentration (see Section 4.10, note 11, p. 45) which is very different from the meaning explained above and should be avoided.

Despite this very common use, it should be noted that the word "molar" violates the rule [3, 5] that neither the name of a quantity, nor the symbol used to denote it, should imply a particular name of a unit ("mole" in this case). The same concern applies to related quantity names, such as "molar mass", "molar volume", "molar gas constant", etc., as well as their recommended symbols involving the letter "M" or "m" as main character or subscript. Other quantity names also violating this rule include "Celsius temperature", "molality" and "mole fraction".

---

[1] The adjective "molar" needs special attention, as described at the end of Section 1.4, where other quantity names needing similar attention are mentioned.

1.5 PRODUCTS AND QUOTIENTS OF PHYSICAL QUANTITIES AND UNITS

Products of physical quantities may be written in any of the ways

$$a\ b \quad \text{or} \quad ab \quad \text{or} \quad a \cdot b \quad \text{or} \quad a \times b$$

When multiplication is indicated by a dot, the dot shall be half high: $a \cdot b$, not $a.b$.
Similarly, quotients may be written

$$a/b \quad \text{or} \quad \frac{a}{b} \quad \text{or by writing the product of } a \text{ and } b^{-1} \text{ as } ab^{-1}$$

Examples $F = ma$, $p = nRT/V$, $c = n\,V^{-1}$

$a : b$ is used for "divided by" as well. However, this symbol is mainly used to express ratios such as length scales in maps.

Not more than one solidus (/) shall be used in the same expression unless parentheses are used to eliminate ambiguity.

Example $(a/b)/c$ or $a/(b/c)$ (in general different), not $a/b/c$

In evaluating combinations of many factors, multiplication written without a multiplication sign takes precedence over division in the sense that a/bc is interpreted as $a/(bc)$ and not as $(a/b)c$; however, it is necessary to use parentheses to eliminate ambiguity under certain circumstances, thus avoiding expressions of the kind a/bcd when ambiguity can arise. Furthermore, $a/b + c$ is interpreted as $(a/b) + c$ and not as $a/(b + c)$.

Products and quotients of units may be written in a similar way, except that the cross (×) is not used as a multiplication sign between units. When a product of units is written without any multiplication sign a space shall be left between the unit symbols.

Example $1\ \mathrm{N} = 1\ \mathrm{m\ kg\ s^{-2}} = 1\ \mathrm{m{\cdot}kg{\cdot}s^{-2}} = 1\ \mathrm{m\ kg/s^2}$, not $1\ \mathrm{mkgs^{-2}}$

1.6 THE USE OF ITALIC AND ROMAN FONTS FOR SYMBOLS IN SCIENTIFIC PUBLICATIONS

Scientific manuscripts should follow the accepted conventions concerning the use of italic and roman fonts for symbols. An italic font is generally used for emphasis in running text, but it has a quite specific meaning when used for symbols in scientific text and equations. The following summary is intended to help in the correct use of italic fonts in preparing manuscript material.

1. The general rules concerning the use of italic (sloping) font or roman (upright) font are presented in Sections 1.3.1 and 1.3.2, p. 3 in relation to symbols for quantities and units, and in Section 1.7, p. 7 in relation to mathematical symbols and operators. These rules are also presented in the International Standards ISO/IEC 80000 [5], ISO 1000 [6], and in the SI Brochure [3].

2. The overall rule is that symbols representing physical quantities or variables are italic, but symbols representing units, mathematical constants, or labels, are roman. Sometimes there may seem to be doubt as to whether a symbol represents a quantity or has some other meaning (such as label): a good rule is that quantities, or variables, may have a range of numerical values, but labels cannot. Vectors, tensors and matrices are denoted using a bold-face (heavy) font, but they shall be italic since they are quantities.

   Examples The electric field strength $\boldsymbol{E}$ has components E_x, E_y, and E_z.
   The mass of my pen is $m = 24\ \mathrm{g} = 0.024\ \mathrm{kg}$.

3. The above rule applies equally to all uppercase and lowercase letter symbols from both the Greek and the Latin alphabet.
   Example When the symbol μ is used to denote a physical quantity (such as permeability or reduced mass) it should be italic, but when it is used as a prefix in a unit such as microgram, μg, or when it is used as the symbol for the muon, μ (see paragraph 5 below), it should be roman.

4. Numbers, and labels, are roman (upright).
   Examples The ground and first excited electronic state of the CH_2 molecule are denoted $\ldots(2a_1)^2(1b_2)^2(3a_1)^1(1b_1)^1$, $\tilde{X}\ ^3B_1$, and $\ldots(2a_1)^2(1b_2)^2(3a_1)^2$, $\tilde{a}\ ^1A_1$, respectively. The π-electron configuration and symmetry of the benzene molecule in its ground state are denoted: $\ldots(a_{2u})^2(e_{1g})^4$, $\tilde{X}\ ^1A_{1g}$. All these symbols are labels and are roman.

5. Symbols for elements in the periodic system shall be roman. Similarly the symbols used to represent elementary particles are always roman. (See, however, paragraph 9 below for use of italic font in chemical-compound names.)
   Examples H, He, Li, Be, B, C, N, O, F, Ne, ... for atoms; e for the electron, p for the proton, n for the neutron, μ for the muon, α for the alpha particle, etc.

6. Symbols for physical quantities are described above in Section 1.3.1.

7. The particular operators ***grad*** and ***rot*** and the corresponding symbols **∇** for ***grad***, **∇**× for ***rot***, and **∇**· for ***div*** are printed in bold-face to indicate the vector or tensor character following [5.b].

8. The fundamental physical constants are always regarded as quantities subject to measurement (even though they are not considered to be variables) and they should accordingly always be italic. Sometimes fundamental physical constants are used as though they were units, but they are still given italic symbols. An example is the hartree, E_h (see Section 3.9.1, p. 13). However, the electronvolt, eV, the dalton, Da, and the unified atomic mass unit, u, have been recognized as units by the Comité International des Poids et Mesures (CIPM) of the BIPM and they are accordingly given roman symbols.
   Examples c_0 for the speed of light in vacuum, m_e for the electron mass, h for the Planck constant, N_A or L for the Avogadro constant, e for the elementary charge, a_0 for the Bohr radius, etc.
   The electronvolt $1\ \mathrm{eV} = e{\cdot}1\ \mathrm{V} = 1.602\ 176\ 634{\times}10^{-19}\ \mathrm{J}$.

9. Greek letters are used in systematic organic, inorganic, macromolecular, and biochemical nomenclature. These should be roman (upright), since they are not symbols for physical quantities. They designate the position of substitution in side chains, ligating-atom attachment and bridging mode in coordination compounds, end groups in structure-based names for macromolecules, and stereochemistry in carbohydrates and natural products. Letter symbols for elements are italic when they are locants in chemical-compound names indicating attachments to heteroatoms, e.g. *O*-, *N*-, *S*-, and *P*-. The italic symbol *H* denotes indicated or added hydrogen (see reference [16]).
   Examples α-ethylcyclopentaneacetic acid
   β-methyl-4-propylcyclohexaneethanol
   α-D-glucopyranose
   N-methylbenzamide
   3*H*-pyrrole

1.7 PRINTING OF NUMBERS AND MATHEMATICAL SYMBOLS [5.a]

1. Numbers in general shall be printed in roman (upright) type. The decimal sign between digits in a number should be a point (e.g. 2.3) or a comma (e.g. 2,3). When the decimal sign is placed before the first significant digit of a number, a zero shall always precede the decimal sign. To facilitate the reading of long numbers the digits may be separated into groups of three about the decimal sign, using only a thin space (but never a point or a comma, nor any other symbol). However, when there are only four digits before or after the decimal marker we recommend that no space is required and no space should be used.

   Examples 2573.421 736 or 2573,421 736 or $0.257\,342\,173\,6\times10^4$ or $0{,}257\,342\,173\,6\times10^4$
   32 573.4215 or 32 573,4215

2. Numerical values of physical quantities which have been experimentally determined are usually subject to some uncertainty. The experimental uncertainty should always be specified. There are several ways to specify such intervals for the uncertainty in the values of quantities.
   The magnitude of the uncertainty may be represented as follows

   Examples $l = [5.3478 - 0.0064, 5.3478 + 0.0064]$ cm
   $l = (5.3478 \pm 0.0064)$ cm
   $l = 5.3478(32)$ cm

   In the first and second examples, the range of expanded uncertainty is indicated directly as $[a-b, a+b]$ and $(a \pm b)$, respectively. It is recommended that these notations should be used only with the meaning that the intervals designated $[a-b, a+b]$ or $(a \pm b)$ contain the true value with a high degree of certainty, such that $b \geq 2u$, where u denotes the standard uncertainty or standard deviation. The uncertainty interval can also be indicated as $a - b \leq l \leq a + b$ (see Chapter 5, p. 67).
   In the third example, $a(c)$, the range of uncertainty c indicated in parentheses is assumed to apply to the least significant digits of a. It is recommended that this notation be reserved for the meaning that c represents $1u$ in the final digits of a [8].
   Nevertheless, when reporting the uncertainty of a measured value, a note must be given explaining what the uncertainty means (see Chapter 5, p. 67).

3. Letter symbols for mathematical constants (e.g. e, π, $\mathrm{i} = \sqrt{-1}$) shall be printed in roman (upright) type, but letter symbols for numbers other than constants (e.g. quantum numbers) should be printed in italic (sloping) type, similar to physical quantities.

4. Symbols for specific mathematical functions and operators (e.g. lb, ln, lg [1], exp, sin, cos, d, δ, Δ, ∇, ...) shall be printed in roman type, but symbols for a general function (e.g. $f(x)$, $F(x, y)$, ...) shall be printed in italic type.

5. The operator p (as in $\mathrm{p}a_{\mathrm{H^+}}$, $\mathrm{p}K = -\lg K$ etc.) shall be printed in roman type.

6. Symbols for symmetry species in group theory (e.g. S, P, D, ..., s, p, d, ..., Σ, Π, Δ, ..., $\mathrm{A_{1g}}$, $\mathrm{B_2''}$, ...) shall be printed in roman (upright) type when they represent the state symbol for an atom or a molecule, although they are often printed in italic type when they represent the symmetry species of a point group.

---

[1] The symbols lb, ln and lg mean $\log_2$, $\log_\mathrm{e}$ and $\log_{10}$. The symbol $\log_a$ defines the logarithm with respect to base a. When the symbol a for the base is omitted there is a risk of ambiguity.

1.8 SYMBOLS, OPERATORS, AND FUNCTIONS [5.b]

Description	*Symbol*	*Notes*
signs and symbols		
equal to	$=$	
not equal to	$\neq$	
identically equal to	$\equiv$	
equal by definition to	$\stackrel{\text{def}}{=}$, $:=$	
approximately equal to	$\approx$	
asymptotically equal to	$\simeq$	
corresponds to	$\widehat{=}$	
proportional to	$\sim, \propto$	
tends to, approaches	$\rightarrow$	
infinity	∞	
less than	$<$	
greater than	$>$	
less than or equal to	$\leqslant$	
greater than or equal to	$\geqslant$	
operations		
mean value of a	$\langle a\rangle$, $\overline{a}$	
sign of a (equal to $a/\lvert a\rvert$ if $a \neq 0$, 0 if $a = 0$)	$\mathrm{sgn}\ a$	
n factorial	$n!$	
binomial coefficient, $n!/p!(n-p)!$	C_p^n, $\binom{n}{p}$	
sum of a_i	$\sum a_i$, $\sum_i a_i$, $\sum\limits_{i=1}^{n} a_i$	
product of a_i	$\prod a_i$, $\prod_i a_i$, $\prod\limits_{i=1}^{n} a_i$	
functions		
greatest integer $\leqslant x$	$\mathrm{ent}\ x$	
integer part of x	$\mathrm{int}\ x$	
integer division	$\mathrm{int}(n/m)$	
square root of -1, $\sqrt{-1}$	i	
change in x	$\Delta x = x(\text{final}) - x(\text{initial})$	
infinitesimal variation of f	δf	
limit of $f(x)$ as x tends to a	$\lim\limits_{x \to a} f(x)$	
1st derivative of f	$\mathrm{d}f/\mathrm{d}x$, f', $(\mathrm{d}/\mathrm{d}x)f$	
nth derivative of f	$\mathrm{d}^n f/\mathrm{d}x^n$, $f^{(n)}$	
partial derivative of f	$\partial f/\partial x$, $\partial_x f$, $\mathrm{D}_x f$	
total differential of f	$\mathrm{d}f$	
inexact differential of f	$\text{đ}f$	1
1st derivative of x with respect to time	$\dot{x}$, $\mathrm{d}x/\mathrm{d}t$	
integral of $f(x)$	$\int f(x)\,\mathrm{d}x$, $\int \mathrm{d}x f(x)$	
Kronecker delta	$\delta_{ij} = \begin{cases} 1 & \text{if } i = j \\ 0 & \text{if } i \neq j \end{cases}$	

(1) Notation used in thermodynamics, see Section 4.11, note 1, p. 49.

Description	*Symbol*	*Notes*
Levi-Civita symbol	$\varepsilon_{ijk} = \begin{cases} 1 & \text{if } ijk \text{ is a cyclic permutation of 123} \\ & \varepsilon_{123} = \varepsilon_{231} = \varepsilon_{312} = 1 \\ -1 & \text{if } ijk \text{ is an anticyclic permutation of 123} \\ & \varepsilon_{132} = \varepsilon_{321} = \varepsilon_{213} = -1 \\ 0 & \text{otherwise} \end{cases}$	
Dirac delta distribution	$\delta(x)$, $\int f(x)\delta(x)\,\mathrm{d}x = f(0)$	
unit step function, Heaviside function	$\varepsilon(x)$, $\mathrm{H}(x)$, $\mathrm{h}(x)$, $\varepsilon(x) = 1$ for $x > 0$, $\varepsilon(x) = 0$ for $x < 0$.	
gamma function	$\Gamma(x) = \int_0^\infty t^{x-1}\mathrm{e}^{-t}\mathrm{d}t$ $\Gamma(n+1) = (n)!$ for positive integers n	
convolution of functions f and g	$f * g = \int_{-\infty}^{+\infty} f(x - x')g(x')\,\mathrm{d}x'$	
vectors		
vector $\boldsymbol{a}$	$\boldsymbol{a}$, $\vec{a}$	
Cartesian components of $\boldsymbol{a}$	a_x, a_y, a_z	
unit vectors (Cartesian)	$\boldsymbol{e}_x$, $\boldsymbol{e}_y$, $\boldsymbol{e}_z$ or $\boldsymbol{i}$, $\boldsymbol{j}$, $\boldsymbol{k}$	
scalar product	$\boldsymbol{a} \cdot \boldsymbol{b}$	
vector or cross product	$\boldsymbol{a} \times \boldsymbol{b}$, $(\boldsymbol{a} \wedge \boldsymbol{b})$	
nabla operator, del operator	$\boldsymbol{\nabla} = \boldsymbol{e}_x\partial/\partial x + \boldsymbol{e}_y\partial/\partial y + \boldsymbol{e}_z\partial/\partial z$	
Laplacian operator	$\boldsymbol{\nabla}^2, \triangle = \partial^2/\partial x^2 + \partial^2/\partial y^2 + \partial^2/\partial z^2$	
gradient of a scalar field V	$\boldsymbol{grad}\ V$, $\boldsymbol{\nabla} V$	
divergence of a vector field $\boldsymbol{A}$	$\boldsymbol{div}\ \boldsymbol{A}$, $\boldsymbol{\nabla} \cdot \boldsymbol{A}$	
rotation of a vector field $\boldsymbol{A}$	$\boldsymbol{rot}\ \boldsymbol{A}$, $\boldsymbol{\nabla} \times \boldsymbol{A}$, $(\boldsymbol{curl}\ \boldsymbol{A})$	
matrices		
matrix of element A_{ij}	$\boldsymbol{A}$	
unit matrix	$\boldsymbol{E}$, $\boldsymbol{I}$	
inverse of a square matrix $\boldsymbol{A}$	$\boldsymbol{A}^{-1}$	
transpose of matrix $\boldsymbol{A}$	$\boldsymbol{A}^{\mathsf{T}}$, $\widetilde{\boldsymbol{A}}$	
complex conjugate of matrix $\boldsymbol{A}$	$\boldsymbol{A}^*$	
conjugate transpose (adjoint) of $\boldsymbol{A}$ (hermitian conjugate of $\boldsymbol{A}$)	$\boldsymbol{A}^{\mathsf{H}}$, $\boldsymbol{A}^\dagger$, where $(\boldsymbol{A}^\dagger)_{ij} = A_{ji}{}^*$	
trace of a square matrix $\boldsymbol{A}$	$\sum_i A_{ii}$, tr $\boldsymbol{A}$	
determinant of a square matrix $\boldsymbol{A}$	det $\boldsymbol{A}$, $\lvert\boldsymbol{A}\rvert$	
sets and logical operators		
p and q (conjunction sign)	$p \wedge q$	
p or q or both (disjunction sign)	$p \vee q$	
negation of p, not p	$\neg p$	
p implies q	$p \Rightarrow q$	
p is equivalent to q	$p \Leftrightarrow q$	
A is contained in B	$A \subset B$	
union of A and B	$A \cup B$	
intersection of A and B	$A \cap B$	
x belongs to A	$x \in A$	
x does not belong to A	$x \notin A$	
the set A contains x	$A \ni x$	
A but not B	$A \backslash B$	

A discussion on units

(paraphrased parody after Umberto Eco 'Il Nome della Rosa' - 'The Name of the Rose')

When Lieutenant Henry Hugh count of Baskerville, also called 'the Hound', towards the end of the war entered with his small group of soldiers this remote Italian mountain village, he addressed its Mayor, asking:

'Tell me, how do the people in the valley make a living?'

'Mostly farming and cattle', was the answer, 'you can assume that a family, in the region of the southern valley owns about 2000 tablets of land.'

'How much is a tablet?'

'Four square trabucci, of course'.

'How much are they?'

'Thirty six square feet is a square trabucco. Or, if you like, eight hundred linear trabbucci make a Piedmont mile. A typical family can cultivate olives for about one sack of oil, and one sack makes five emine and one emina makes eight cups.'

'I hope to understand', said the Hound with some bewilderment, 'Every country has its own measures. For instance, while the British measure beer by the pint, you measure wine with the boccali.'

'Or with the rubbio. Six rubbie make one brenta and eight brente make one bottale. If you like, one rubbio gives six pints. You should note, however, that our pint is different from the British imperial pint, which is about ten percent larger than 500 cm^3, whereas the American liquid pint is a little less than 500 cm^3 , if you want to translate this into metric units.'

'I think that I get the idea quite clearly', said the Hound with some resignation.

'Yes , indeed', said the Mayor, 'for instance in "New" England (USA) you still use the inch to measure length and the pound per square inch, the psi, to measure pressure. Do you wish to know anything else?', he added with a tone that seemed defiant.

'No, thank you, that is more than enough', the Lieutenant concluded this discussion.

2 DEFINITIONS AND SYMBOLS FOR UNITS

2.1 THE INTERNATIONAL SYSTEM OF UNITS (SI)

The International System of Units (SI) was adopted in 1960 [3]. The SI is set up on the basis of seven *base units*, which are given in Section 2.2. The *SI derived units* are expressed as products of the base units. As of May 20, 2019, the SI is defined by fixing the numerical values of seven physical constants [3].

The SI is the system of units in which

the unperturbed ground state hyperfine transition frequency of the caesium-133 atom $\Delta\nu_{\mathrm{Cs}}$ is 9 192 631 770 Hz,

the speed of light in vacuum c is 299 792 458 m/s,

the Planck constant h is $6.626\,070\,15\times10^{-34}$ J s,

the elementary charge e is $1.602\,176\,634\times10^{-19}$ C,

the Boltzmann constant k is $1.380\,649\times10^{-23}$ J/K,

the Avogadro constant N_{A} is $6.022\,140\,76\times10^{23}$ mol^{-1},

the luminous efficacy of monochromatic radiation of frequency 540×10^{12} Hz, K_{cd}, is 683 lm/W.

Some of these physical constants are also given in the abbreviated list of fundamental constants in the cover page of this book.

In the International System of Units there is only one coherent SI unit for each physical quantity. This is either the appropriate SI base unit itself or the appropriate SI derived unit. However, any of the approved decimal prefixes, called *SI prefixes*, may be used to construct decimal multiples or submultiples of SI units (see Section 2.5 below). It is recommended that units of the SI be used in science and technology (for more details, see [3]). When other units are used, they should be clearly defined in relation to the SI units.

2.2 NAMES AND SYMBOLS FOR THE SI BASE UNITS

The symbols listed here are internationally agreed and shall not be changed in other languages or scripts. See sections 1.3.2 and 1.6, p. 3 and p. 5 on the printing of symbols for units.

	SI base unit	
Base quantity	*Name*	*Symbol*
length	metre	m
mass	kilogram	kg
time	second	s
electric current	ampere	A
thermodynamic temperature	kelvin	K
amount of substance	mole	mol
luminous intensity	candela	cd

2.3 COHERENT UNITS AND CHECKING DIMENSIONS

If equations between numerical values have the same form as equations between physical quantities, then the system of units defined in terms of base units avoids numerical factors between units, and is said to be a *coherent system*. For example, the kinetic energy T of a particle of mass m moving with a speed v is defined by the equation

$$T = (1/2)\ mv^2$$

but the SI unit of kinetic energy is the joule, defined by the equation

$$\mathrm{J} = \mathrm{kg}\ (\mathrm{m/s})^2 = \mathrm{kg\ m^2\ s^{-2}}$$

where it is to be noted that the factor (1/2) is omitted. In fact the joule is simply a special name and its symbol J stands for the product of units kg $\mathrm{m^2\ s^{-2}}$.

The International System (SI) is a coherent system of units. The advantage of a coherent system of units is that if the value of each quantity is substituted for the quantity symbol in any quantity equation, then the units may be canceled, leaving an equation between numerical values which is exactly similar (including all numerical factors) to the original equation between the quantities. Checking that the units cancel in this way is sometimes described as checking the dimensions of the equation.

The use of a coherent system of units is not essential. In particular the use of multiple or submultiple prefixes destroys the coherence of the SI, but is nonetheless often convenient.

2.4 PHYSICAL CONSTANTS USED AS ATOMIC UNITS

Sometimes fundamental physical constants, or other well defined physical quantities, are used as though they were units in certain specialized fields of science. For example, in astronomy it may be more convenient to express the mass of a star in terms of the mass of the sun. In atomic and molecular physics it is similarly more convenient to express masses in terms of the electron mass, m_e, or in terms of the unified atomic mass unit, 1 u, and to express charges in terms of the elementary charge e, and energies in terms of the electronvolt, eV.

The electronvolt is the kinetic energy acquired by an electron in passing through a potential difference of 1 V in vacuum, 1 eV = $1.602\ 176\ 634\times10^{-19}$ J. The numerical value of a quantity expressed in this unit may be converted into its value when expressed in the SI by multiplication with the value of the physical constant in the SI.

The dalton and the unified atomic mass unit are alternative names for the same unit, therefore 1 u = 1 Da $\approx 1.6605\times10^{-27}$ kg. The dalton may be combined with the SI prefixes to express the masses of large molecules in kilodalton (kDa) or megadalton (MDa).

Physical quantity	*Physical constant*	*Symbol for unit*	*Value in SI units*	*Notes*
mass	electron mass	m_e	$\approx 9.1094\times10^{-31}$ kg	
charge	elementary charge	e	$= 1.602\ 176\ 634\times10^{-19}$ C	
action, (angular momentum)	Planck constant divided by 2π	$\hbar$	$= 1.054\ 571\ 817...\times10^{-34}$ J s	1
length	bohr	a_0	$\approx 5.2918\times10^{-11}$ m	1
energy	hartree	E_h	$\approx 4.3597\times10^{-18}$ J	1
time		$\hbar/E_h$	$\approx 2.4189\times10^{-17}$ s	
speed		$a_0E_h/\hbar$	$\approx 2.1877\times10^{6}$ m s^{-1}	2
electric field strength		E_h/ea_0	$\approx 5.1422\times10^{11}$ V m^{-1}	
electric dipole moment		ea_0	$\approx 8.4784\times10^{-30}$ C m	
electric quadrupole moment		$ea_0{}^2$	$\approx 4.4866\times10^{-40}$ C m^2	
electric polarizability		$e^2a_0{}^2/E_h$	$\approx 1.6488\times10^{-41}$ C^2 m^2 J^{-1}	
magnetic flux density		$\hbar/ea_0{}^2$	$\approx 2.3505\times10^{5}$ T	
magnetic dipole moment		$e\hbar/m_e$	$\approx 1.8548\times10^{-23}$ J T^{-1}	3

(1) $\hbar = h/2\pi$; $a_0 = 4\pi\varepsilon_0\hbar^2/m_ee^2$; $E_h = \hbar^2/m_ea_0{}^2$.
(2) The numerical value of the speed of light, when expressed in atomic units, is equal to the reciprocal of the fine-structure constant α; $c/(a_0E_h/\hbar) = c\hbar/a_0E_h = \alpha^{-1} = 137.035\,999\,084(21)$.
(3) The atomic unit of magnetic dipole moment is twice the Bohr magneton, μ_B.

One particular group of physical constants that are used as though they were units deserve special mention. These are the so-called *atomic units* and arise in calculations of electronic wavefunctions for atoms and molecules, i.e. in quantum chemistry. Only the first five atomic units in the table above have special names and symbols.

The relation of atomic units to the corresponding SI units involves the values of the fundamental physical constants, and is therefore not exact. The numerical values in the table are rounded from the CODATA compilation [17, 18]. The numerical results of calculations in theoretical chemistry are frequently quoted in atomic units, or as numerical values in the form *physical quantity* divided by *atomic unit*, so that the reader may make the conversion using the current best estimates of the physical constants.

Many authors make no use of the symbols for the atomic units listed in the tables above, but instead use the symbol "a.u." or "au" for all atomic units. This custom should not be followed. It leads to confusion, just as it would if we were to write "SI" as a symbol for every SI unit, or "cgs" as a symbol for every cgs unit (the 'centimetre, gram, second' system of units, see Section 3.2, p. 16).

Examples For the hydrogen molecule the equilibrium bond length r_e, and the dissociation energy D_e, are given by

$r_e = 2.1\ a_0$ *not* $r_e = 2.1$ a.u.

$D_e = 0.16\ E_h$ *not* $D_e = 0.16$ a.u.

2.5 SI PREFIXES AND PREFIXES FOR BINARY MULTIPLES

The following prefixes [3] are used to denote decimal multiples and submultiples of SI units.

	Prefix			*Prefix*	
Submultiple	*Name*	*Symbol*	*Multiple*	*Name*	*Symbol*
10^{-1}	deci	d	10^{1}	deca	da
10^{-2}	centi	c	10^{2}	hecto	h
10^{-3}	milli	m	10^{3}	kilo	k
10^{-6}	micro	µ	10^{6}	mega	M
10^{-9}	nano	n	10^{9}	giga	G
10^{-12}	pico	p	10^{12}	tera	T
10^{-15}	femto	f	10^{15}	peta	P
10^{-18}	atto	a	10^{18}	exa	E
10^{-21}	zepto	z	10^{21}	zetta	Z
10^{-24}	yocto	y	10^{24}	yotta	Y

Prefix symbols shall be printed in roman (upright) type with no space between the prefix and the unit symbol.

Example kilometre, km

When a prefix is used with a unit symbol, the combination is taken as a new symbol that can be raised to any power without the use of parentheses.

Examples $1\ \mathrm{cm}^3 = (10^{-2}\ \mathrm{m})^3 = 10^{-6}\ \mathrm{m}^3$

A prefix shall never be used on its own, and prefixes are not to be combined into compound prefixes.

Example pm, not µµm

The names and symbols of decimal multiples and submultiples of the SI base unit of mass, the kilogram, symbol kg, which already contains a prefix, are constructed by adding the appropriate prefix to the name gram and symbol g.

Examples mg, not µkg; Mg, not kkg

The International Electrotechnical Commission (IEC) has standardized the following prefixes for binary multiples, mainly used in information technology, to be distinguished from the SI prefixes for decimal multiples [7].

	Prefix		
Multiple	*Name*	*Symbol*	*Origin*
$(2^{10})^1 = (1024)^1$	kibi	Ki	kilobinary
$(2^{10})^2 = (1024)^2$	mebi	Mi	megabinary
$(2^{10})^3 = (1024)^3$	gibi	Gi	gigabinary
$(2^{10})^4 = (1024)^4$	tebi	Ti	terabinary
$(2^{10})^5 = (1024)^5$	pebi	Pi	petabinary
$(2^{10})^6 = (1024)^6$	exbi	Ei	exabinary
$(2^{10})^7 = (1024)^7$	zebi	Zi	zettabinary
$(2^{10})^8 = (1024)^8$	yobi	Yi	yottabinary

3 CONVERSION OF UNITS

Units of the SI are recommended for use throughout science and technology. However, the published literature of science makes widespread use of non-SI units, e.g. Torr, bar, atm, kcal. It is thus often necessary to convert the values of physical quantities between SI units and other units. This chapter is concerned with facilitating this process.

3.1 THE USE OF QUANTITY CALCULUS

Quantity calculus is a system of algebra in which symbols are consistently used to represent physical quantities as the product of a numerical value and a unit and in which we manipulate the symbols for physical quantities, numerical values, and units by the ordinary rules of algebra. Quantity calculus has particular advantages in facilitating the problems of converting between different units and different systems of units. This is demonstrated in the following examples using a limited number of significant digits.

Example 1. The vapor pressure of water at 20 °C is recorded to be

$$p(\mathrm{H_2O},\ 20\ °\mathrm{C}) \approx 17.5\ \mathrm{Torr}$$

The torr, the bar, and the atmosphere are given by the equations

$$1\ \mathrm{Torr} \approx 133.3\ \mathrm{Pa},\ 1\ \mathrm{bar} = 10^5\ \mathrm{Pa},\ \text{and}\ 1\ \mathrm{atm} = 101\ 325\ \mathrm{Pa}$$

Thus

$$\begin{aligned} p(\mathrm{H_2O},\ 20\ °\mathrm{C}) &\approx 17.5{\times}133.3\ \mathrm{Pa} \approx 2.33\ \mathrm{kPa} = \\ &(2.33{\times}10^3/10^5)\ \mathrm{bar} = 23.3\ \mathrm{mbar} = \\ &(2.33{\times}10^3/101\ 325)\ \mathrm{atm} \approx 2.30{\times}10^{-2}\ \mathrm{atm} \end{aligned}$$

Example 2. Spectroscopic measurements show that for the methylene radical, $\mathrm{CH_2}$, the $\widetilde{\mathrm{a}}\ ^1\mathrm{A_1}$ excited state lies at a repetency (wavenumber) 3156 $\mathrm{cm^{-1}}$ above the $\widetilde{\mathrm{X}}\ ^3\mathrm{B_1}$ ground state

$$\widetilde{\nu}(\widetilde{\mathrm{a}} - \widetilde{\mathrm{X}}) = T_0(\widetilde{\mathrm{a}}) - T_0(\widetilde{\mathrm{X}}) \approx 3156\ \mathrm{cm^{-1}}$$

The excitation energy from the ground triplet state to the excited singlet state is thus

$$\begin{aligned} \Delta E = hc_0\widetilde{\nu} &\approx (6.626{\times}10^{-34}\ \mathrm{J\ s})\ (2.998{\times}10^8\ \mathrm{m\ s^{-1}})\ (3156\ \mathrm{cm^{-1}}) \approx \\ &6.269{\times}10^{-22}\ \mathrm{J\ m\ cm^{-1}} = 6.269{\times}10^{-20}\ \mathrm{J} = 6.269{\times}10^{-2}\ \mathrm{aJ} \end{aligned}$$

where the values of h and c_0 are taken from the fundamental physical constants (see cover page) and we have used the relation 1 m = 100 cm, or 1 m 1 $\mathrm{cm^{-1}}$ = 100. Since the electronvolt is given by the equation 1 eV ≈ $1.6022{\times}10^{-19}$ J, or 1 aJ ≈ (1/0.160 22) eV,

$$\Delta E \approx \left(6.269 \times 10^{-2}/0.160\ 22\right)\ \mathrm{eV} \approx 0.3913\ \mathrm{eV}$$

Similarly the hartree is given by $E_\mathrm{h} = \hbar^2/m_\mathrm{e}a_0{}^2 \approx 4.3597$ aJ, or 1 aJ ≈ $(1/4.3597)E_\mathrm{h}$, and thus the excitation energy is given in atomic units by

$$\Delta E \approx \left(6.269 \times 10^{-2}/4.3597\right)\ E_\mathrm{h} \approx 1.438 \times 10^{-2}\, E_\mathrm{h}$$

Finally the molar excitation energy is given by

$$\Delta E_\mathrm{m} = N_\mathrm{A}\Delta E \approx (6.022 \times 10^{23}\ \mathrm{mol^{-1}})(6.269 \times 10^{-2}\ \mathrm{aJ}) \approx 37.75\ \mathrm{kJ\ mol^{-1}}$$

Also, since 1 kcal = 4.184 kJ, or 1 kJ = (1/4.184) kcal,

$$\Delta E_\mathrm{m} \approx (37.75/4.184)\ \mathrm{kcal\ mol^{-1}} \approx 9.023\ \mathrm{kcal\ mol^{-1}}$$

In the transformation from ΔE to ΔE_m the coefficient N_A (Avogadro constant), is not a number, but has the unit $\mathrm{mol^{-1}}$.

3.2 CONVERSION TABLES FOR UNITS

The table below gives conversion factors from a variety of units to the corresponding SI unit [17]. For each physical quantity the name is given, followed by the recommended symbol(s), the SI unit name and its symbol. Entries give other units in common use, with their conversion factors to SI and other units. Systems of units other than the SI are referred to by the following acronyms: au (atomic units), cgs ('centimetre, gram, second' system of units), esu (electrostatic system of units), emu (electromagnetic system of units), and Gaussian (Gaussian system of units). The constant ζ which occurs in some of the electromagnetic conversion factors is the exact number 29 979 245 800 and equals $c_0/(\text{cm s}^{-1})$.

The inclusion of non-SI units in this table should not be taken to imply that their use is to be encouraged. With some exceptions, SI units are always to be preferred to non-SI units. However, since many of the units below are to be found in the scientific literature, it is convenient to tabulate their relation to the SI.

The table must be read in the following way, for example for the ångström, as a unit of length: $\text{Å} = 10^{-10}$ m means that the symbol Å of a length 13.1 Å may be replaced by 10^{-10} m, saying that the length has the value 13.1×10^{-10} m. Conversion factors are either given exactly or rounded to four digits (when the sign $\approx$ is used) for convenience only. An entry in the column named "*Symbol*" may refer to unit or quantity symbol (see Sections 1.3.1 and 1.3.2, p. 2).

Name (Unit system)	*Symbol*	*Expressed in SI units*	*Notes*
length, l, metre, m			
bohr	a_0	$= 4\pi\epsilon_0\hbar^2/m_e e^2 \approx 5.2918\times10^{-11}$ m	
ångström	Å	$= 10^{-10}$ m	
micron	µ	$= 1$ µm $= 10^{-6}$ m	
inch	in	$= 2.54\times10^{-2}$ m	
foot	ft	= 12 in = 0.3048 m	
yard	yd	= 3 ft = 0.9144 m	
mile	mi	= 1760 yd = 1609.344 m	
area, A, square metre, m^2			
barn	b	$= 10^{-28}$ m^2	
volume, V, cubic metre, m^3			
litre	l, L	$= 1$ dm^3 $= 10^{-3}$ m^3	
plane angle, α, radian, rad			
degree	°, deg	$= (\pi/180)$ rad $\approx (1/57.295\ 78)$ rad	
mass, m, kilogram, kg			
electron mass	m_e	$\approx 9.1094\times10^{-31}$ kg	
dalton, unified atomic mass unit	Da, u	$= m_a(^{12}C)/12 \approx 1.6605\times10^{-27}$ kg	
time, t, second, s			
au of time	$\hbar/E_h$	$\approx 2.4189\times10^{-17}$ s	
minute	min	= 60 s	
hour	h	= 3600 s	
day	d	= 24 h = 86 400 s	
year	a	= 31 556 925.9747 s	1

(1) This number refers to the tropical year for the epoch 1900.0. For more details, see [19] and the 4th edition of the Green Book [20].

Name (Unit system)	*Symbol*	*Expressed in SI units*	*Notes*
speed, v, metre per second, $m\ s^{-1}$			
au of speed	$a_0 E_h/\hbar$	$\approx 2.1877\times10^6\ m\ s^{-1}$	
knot	kn	$= 1.852\ m\ s^{-1}$	
acceleration, a, $m\ s^{-2}$			
standard acceleration of gravity	g_n	$= 9.806\ 65\ m\ s^{-2}$	
force, F, newton, $N = 1\ kg\ m\ s^{-2}$			
dyne (cgs unit)	dyn	$= 1\ g\ cm\ s^{-2} = 10^{-5}\ N$	
au of force	E_h/a_0	$\approx 8.2387\times10^{-8}\ N$	
kilogram-force, kilopond	kgf = kp	$= 9.806\ 65\ N$	
energy, E, U, joule, $J = 1\ kg\ m^2\ s^{-2}$			
erg (cgs unit)	erg	$= 1\ g\ cm^2\ s^{-2} = 10^{-7}\ J$	
hartree	E_h	$= \hbar^2/m_e a_0{}^2 \approx 4.3597\times10^{-18}\ J$	
rydberg	Ry	$= E_h/2 \approx 2.1799\times10^{-18}\ J$	
electronvolt	eV	$= e{\cdot}1\ V \approx 1.6022\times10^{-19}\ J$	
calorie, thermochemical	cal_{th}, cal	$= 4.184\ J$	
calorie, international	cal_{IT}	$= 4.1868\ J$	
15 °C calorie	cal_{15}	$\approx 4.1855\ J$	
kilowatt-hour	kWh	$= 3.6\times10^6\ J$	
pressure, p, pascal, $Pa = 1\ N\ m^{-2} = 1\ kg\ m^{-1}\ s^{-2}$			
standard atmosphere	atm	$= 101\ 325\ Pa$	
bar	bar	$= 10^5\ Pa$	
torr	Torr	$= (101\ 325/760)\ Pa \approx 133.322\ Pa$	
conventional millimetre Hg	mmHg	$\approx 133.322\ Pa$	
power, P, watt, $W = 1\ kg\ m^2\ s^{-3}$			
metric horse power	hk	$= 735.498\ 75\ W$	
angular momentum, L, J, $J\ s = 1\ kg\ m^2\ s^{-1}$			
action (cgs unit)	erg s	$= 10^{-7}\ J\ s$	
au of action	$\hbar$	$= h/2\pi \approx 1.0546\times10^{-34}\ J\ s$	
dynamic viscosity, η, $Pa\ s = 1\ kg\ m^{-1}\ s^{-1}$			
poise (cgs unit)	P	$= 10^{-1}\ Pa\ s$	
thermodynamic temperature, T, kelvin, K			
degree Rankine	°R	$= (5/9)\ K$	2
Celsius temperature, t, degree Celsius, °C			2, 3
activity, A, becquerel, $Bq = 1\ s^{-1}$			
curie	Ci	$= 3.7\times10^{10}\ Bq$	
absorbed dose of radiation, D, gray, $Gy = 1\ J\ kg^{-1} = 1\ m^2\ s^{-2}$			
dose equivalent, H, sievert, $Sv = 1\ J\ kg^{-1} = 1\ m^2\ s^{-2}$			

(2) The Rankine temperature, the Celsius temperature t, and the Fahrenheit temperature t_F are related to the thermodynamic temperature T as follows: $T/°R = (9/5)T/K$ (Rankine), $t/°C = T/K - 273.15$ (Celsius), $t_F/°F = (9/5)(t/°C) + 32$ (Fahrenheit).

(3) See also footnote 2, p. 4.

Name (Unit system)	*Symbol*	*Expressed in SI units*	*Notes*
electric current, I, ampere, A			
(esu, Gaussian)	$\mathrm{Fr\ s^{-1}}$	$= (10/\zeta)\mathrm{A} \approx 3.3356\times10^{-10}\ \mathrm{A}$	4
biot (emu)	Bi	$= 10\ \mathrm{A}$	
electric charge, Q, coulomb, C = 1 A s			
franklin (esu, Gaussian)	Fr	$= (10/\zeta)\mathrm{C} \approx 3.3356\times10^{-10}\ \mathrm{C}$	4
(emu)	Bi s	$= 10\ \mathrm{C}$	
proton charge (au)	e	$= 1.602\ 176\ 634\times10^{-19}\ \mathrm{C}$	
charge density, ρ, $\mathrm{C\ m^{-3}} = 1\ \mathrm{A\ s\ m^{-3}}$			
(esu, Gaussian)	$\mathrm{Fr\ cm^{-3}}$	$\approx 3.3356\times10^{-4}\ \mathrm{C\ m^{-3}}$	
(au)	$e{a_0}^{-3}$	$\approx 1.0812\times10^{12}\ \mathrm{C\ m^{-3}}$	
electric potential, V, φ, volt, $\mathrm{V} = 1\ \mathrm{J\ C^{-1}} = 1\ \mathrm{kg\ m^2\ s^{-3}\ A^{-1}}$			
(esu, Gaussian)	$\mathrm{Fr\ cm^{-1}}$	$= \zeta\times10^{-8}\ \mathrm{V} \approx 299.7925\ \mathrm{V}$	4
(au)	$e/4\pi\varepsilon_0 a_0$	$= E_\mathrm{h}/e \approx 27.2114\ \mathrm{V}$	
electric resistance, R, ohm, $\Omega = 1\ \mathrm{V\ A^{-1}} = 1\ \mathrm{m^2\ kg\ s^{-3}\ A^{-2}}$			
(Gaussian)	$\mathrm{s\ cm^{-1}}$	$= \zeta^2\times10^{-9}\ \Omega \approx 8.9876\times10^{11}\ \Omega$	4
conductivity, κ, σ, $\mathrm{S\ m^{-1}} = 1\ \mathrm{kg^{-1}\ m^{-3}\ s^3\ A^2}$			
(Gaussian)	$\mathrm{s^{-1}}$	$= (10^{11}/\zeta^2)\ \mathrm{S\ m^{-1}} \approx$ $1.1127\times10^{-10}\ \mathrm{S\ m^{-1}}$	
capacitance, C, farad, $\mathrm{F} = 1\ \mathrm{kg^{-1}\ m^{-2}\ s^4\ A^2}$			
(Gaussian)	cm	$= (10^9/\zeta^2)\ \mathrm{F} \approx 1.1127\times10^{-12}\ \mathrm{F}$	4
electric field strength, E, $\mathrm{V\ m^{-1}} = 1\ \mathrm{J\ C^{-1}\ m^{-1}} = 1\ \mathrm{kg\ m\ s^{-3}\ A^{-1}}$			
(esu, Gaussian)	$\mathrm{Fr^{-1}\ cm^{-2}}$	$= \zeta\times10^{-6}\ \mathrm{V\ m^{-1}} \approx 2.9979\times10^4\ \mathrm{V\ m^{-1}}$	
(au)	E_h/ea_0	$\approx 5.1422\times10^{11}\ \mathrm{V\ m^{-1}}$	
electric dipole moment, p, μ, C m = 1 A s m			
debye	D	$\approx 3.3356\times10^{-30}\ \mathrm{C\ m}$	
(au)	ea_0	$\approx 8.4784\times10^{-30}\ \mathrm{C\ m}$	
electric quadrupole moment, $Q_{\alpha\beta}$, $\Theta_{\alpha\beta}$, eQ, $\mathrm{C\ m^2} = 1\ \mathrm{A\ s\ m^2}$			
(esu, Gaussian)	$\mathrm{Fr\ cm^2}$	$\approx 3.3356\times10^{-14}\ \mathrm{C\ m^2}$	
(au)	$e{a_0}^2$	$\approx 4.4866\times10^{-40}\ \mathrm{C\ m^2}$	
polarizability, α, $\mathrm{J^{-1}\ C^2\ m^2} = 1\ \mathrm{F\ m^2} = 1\ \mathrm{kg^{-1}\ s^4\ A^2}$			
(esu, Gaussian)	$\mathrm{Fr^2\ cm^3}$	$= (10^5/\zeta^2)\ \mathrm{J^{-1}\ C^2\ m^2} \approx$ $1.1127\times10^{-16}\ \mathrm{J^{-1}\ C^2\ m^2}$	4
"cm³"	$4\pi\varepsilon_0\ \mathrm{cm^3}$	$\approx 1.1127\times10^{-16}\ \mathrm{J^{-1}\ C^2\ m^2}$	5
(au)	$e^2{a_0}^2/E_\mathrm{h}$	$\approx 1.6488\times10^{-41}\ \mathrm{J^{-1}\ C^2\ m^2}$	

(4) ζ is the exact number $\zeta = c_0/(\mathrm{cm\ s^{-1}}) = 29\ 979\ 245\ 800$.
(5) The unit in quotation marks for polarizability ('polarizability volume') may be found in the literature, although it is formally incorrect.

Name (Unit system)	*Symbol*	*Expressed in SI units*	*Notes*
electric displacement, D, (volume) polarization, P, $\mathrm{C\ m^{-2} = 1\ A\ s\ m^{-2}}$			
(esu, Gaussian)	$\mathrm{Fr\ cm^{-2}}$	$= (10^5/\zeta)\ \mathrm{C\ m^{-2}} \approx 3.3356\times10^{-6}\ \mathrm{C\ m^{-2}}$	4
magnetic flux density, B, (magnetic field), tesla, $\mathrm{T = 1\ J\ A^{-1}\ m^{-2} = 1\ V\ s\ m^{-2} = 1\ Wb\ m^{-2}}$			
gauss (emu, Gaussian)	G	$= 10^{-4}\ \mathrm{T}$	
(au)	$\hbar/ea_0{}^2$	$\approx 2.3505\times10^5\ \mathrm{T}$	
magnetic flux, Φ, weber, $\mathrm{Wb = 1\ J\ A^{-1} = 1\ V\ s = 1\ kg\ m^2\ s^{-2}\ A^{-1}}$			
maxwell (emu, Gaussian)	Mx	$= 10^{-8}\ \mathrm{Wb}\ [= 1\ \mathrm{G\ cm^2}]$	
magnetic field strength, H, $\mathrm{A\ m^{-1}}$			
oersted (emu, Gaussian)	Oe	$= (10^3/4\pi)\ \mathrm{A\ m^{-1}}$	6
(volume) magnetization, M, $\mathrm{A\ m^{-1}}$			
gauss (emu, Gaussian)	G	$= 10^3\ \mathrm{A\ m^{-1}}$	6
magnetic dipole moment, m, μ, $\mathrm{J\ T^{-1} = 1\ A\ m^2}$			
(emu, Gaussian)	$\mathrm{erg\ G^{-1}}$	$= 10\ \mathrm{A\ cm^2} = 10^{-3}\ \mathrm{J\ T^{-1}}$	
Bohr magneton	μ_B	$= e\hbar/2m_\mathrm{e} \approx 9.2740\times10^{-24}\ \mathrm{J\ T^{-1}}$	
(au)	$e\hbar/m_\mathrm{e}$	$= 2\mu_\mathrm{B} \approx 1.8548\times10^{-23}\ \mathrm{J\ T^{-1}}$	
nuclear magneton	μ_N	$= (m_\mathrm{e}/m_\mathrm{p})\mu_\mathrm{B} \approx 5.0508\times10^{-27}\ \mathrm{J\ T^{-1}}$	
magnetizability, ξ, $\mathrm{J\ T^{-2} = 1\ A^2\ s^2\ m^2\ kg^{-1}}$			
(Gaussian)	$\mathrm{erg\ G^{-2}}$	$= 10\ \mathrm{J\ T^{-2}}$	
(au)	$e^2a_0{}^2/m_\mathrm{e}$	$\approx 7.8910\times10^{-29}\ \mathrm{J\ T^{-2}}$	
magnetic susceptibility, χ, κ, 1			
(emu, Gaussian)	1		7
molar magnetic susceptibility, χ_m, $\mathrm{m^3\ mol^{-1}}$			
(emu, Gaussian)	$\mathrm{cm^3\ mol^{-1}}$	$= 10^{-6}\ \mathrm{m^3\ mol^{-1}}$	3, 7, 8
inductance, self-inductance, L, henry, $\mathrm{H = 1\ V\ s\ A^{-1} = 1\ kg\ m^2\ s^{-2}\ A^{-2}}$			
(Gaussian)	$\mathrm{s^2\ cm^{-1}}$	$= \zeta^2 \times 10^{-9}\ \mathrm{H} \approx 8.9876\times10^{11}\ \mathrm{H}$	4
(emu)	cm	$= 10^{-9}\ \mathrm{H}$	

(6) In practice the oersted, Oe, is only used as a unit for $H^{(\mathrm{ir})} = 4\pi H$, thus when $H^{(\mathrm{ir})} = 1$ Oe, $H = (10^3/4\pi)$ $\mathrm{A\ m^{-1}}$ (see Section 7.3, ref. [20]). In the Gaussian or emu system, gauss and oersted are equivalent units.
(7) In practice susceptibilities quoted in the context of emu or Gaussian units are always values for $\chi^{(\mathrm{ir})} = \chi/4\pi$; thus when $\chi^{(\mathrm{ir})} = 10^{-6}$, $\chi = 4\pi\times10^{-6}$ (see Section 7.3, ref. [20]).
(8) In practice the units $\mathrm{cm^3\ mol^{-1}}$ usually imply that the non-rationalized molar susceptibility is being quoted $\chi_\mathrm{m}{}^{(\mathrm{ir})} = \chi_\mathrm{m}/4\pi$. For example if $\chi_\mathrm{m}{}^{(\mathrm{ir})} = -15\times10^{-6}\ \mathrm{cm^3\ mol^{-1}}$, then $\chi_\mathrm{m} = -1.88\times10^{-10}\ \mathrm{m^3\ mol^{-1}}$ (see Section 7.3, ref. [20]).

3.3 TRANSFORMATION OF EQUATIONS OF ELECTROMAGNETIC THEORY BETWEEN THE ISQ (SI) AND GAUSSIAN FORMS

Transformation of other important equations of electromagnetic theory are given in the Green Book, 3rd Edition (2011) [2.c] and 4th Edition [20]. There, the reader will also find general forms of these equations that allow writing them using other systems of units, such as the system of atomic units.

	ISQ (SI)	*Gaussian*[1]
Field due to a charge distribution ρ in vacuum (Gauss law):		
	$\nabla \cdot \boldsymbol{E} = \rho/\varepsilon_0$	$\nabla \cdot \boldsymbol{E} = 4\pi\rho$
Potential around a dipole in vacuum:		
	$\phi = \boldsymbol{p} \cdot \boldsymbol{r}/4\pi\varepsilon_0 r^3$	$\phi = \boldsymbol{p} \cdot \boldsymbol{r}/r^3$
Dielectric polarization:		
	$\boldsymbol{P} = \chi_e \varepsilon_0 \boldsymbol{E}$	$\boldsymbol{P} = {\chi_e}^{(ir)} \boldsymbol{E}$
Electric susceptibility and relative permittivity:		
	$\varepsilon_r = 1 + \chi_e$	$\varepsilon_r = 1 + {\chi_e}^{(ir)}$
Electric displacement:		
	$\boldsymbol{D} = \varepsilon_0 \boldsymbol{E} + \boldsymbol{P}$	$\boldsymbol{D}^{(ir)} = \boldsymbol{E} + 4\pi\boldsymbol{P}$
for isotropic media:	$\boldsymbol{D} = \varepsilon_0 \varepsilon_r \boldsymbol{E}$	$\boldsymbol{D}^{(ir)} = \varepsilon_r \boldsymbol{E}$
Magnetic flux density due to a current density $\boldsymbol{j}$ in vacuum (Ampère law):		
	$\nabla \times \boldsymbol{B} = \mu_0 \boldsymbol{j}$	$\nabla \times \boldsymbol{B} = 4\pi\boldsymbol{j}/c_0$
Magnetic dipole of a current loop of area $\boldsymbol{A}$:		
	$\boldsymbol{m} = I\boldsymbol{A}/c_0$	
Magnetization:		
	$\boldsymbol{M} = \chi \boldsymbol{H}$	$\boldsymbol{M} = \chi^{(ir)} \boldsymbol{H}^{(ir)}$
Magnetic susceptibility and relative permeability:		
	$\mu_r = 1 + \chi$	$\mu_r = 1 + 4\pi\chi^{(ir)}$
Magnetic field strength:		
	$\boldsymbol{H} = \boldsymbol{B}/\mu_0 - \boldsymbol{M}$	$\boldsymbol{H}^{(ir)} = \boldsymbol{B} - 4\pi\boldsymbol{M}$
for isotropic media:	$\boldsymbol{H} = \boldsymbol{B}/\mu_0 \mu_r$	$\boldsymbol{H}^{(ir)} = \boldsymbol{B}/\mu_r$
Conductivity:		
$\boldsymbol{j} = \kappa \boldsymbol{E}$	$\boldsymbol{j} = \kappa \boldsymbol{E}$	$\boldsymbol{j} = \kappa \boldsymbol{E}$
Faraday induction law:		
$\nabla \times \boldsymbol{E} + \frac{1}{k}\frac{\partial \boldsymbol{B}}{\partial t} = \boldsymbol{0}$	$\nabla \times \boldsymbol{E} + \partial \boldsymbol{B}/\partial t = \boldsymbol{0}$	$\nabla \times \boldsymbol{E} + \frac{1}{c_0}\frac{\partial \boldsymbol{B}}{\partial t} = \boldsymbol{0}$
Relation between the electric field strength and electromagnetic potentials:		
$\boldsymbol{E} = -\nabla\phi - \frac{1}{k}\frac{\partial \boldsymbol{A}}{\partial t}$	$\boldsymbol{E} = -\nabla\phi - \partial \boldsymbol{A}/\partial t$	$\boldsymbol{E} = -\nabla\phi - \frac{1}{c_0}\frac{\partial \boldsymbol{A}}{\partial t}$
Maxwell equations:		
	$\nabla \cdot \boldsymbol{D} = \rho$	$\nabla \cdot \boldsymbol{D}^{(ir)} = 4\pi\rho$
	$\nabla \times \boldsymbol{H} - \partial \boldsymbol{D}/\partial t = \boldsymbol{j}$	$\nabla \times \boldsymbol{H}^{(ir)} - \frac{1}{c_0}\frac{\partial \boldsymbol{D}^{(ir)}}{\partial t} = \frac{4\pi}{c_0}\boldsymbol{j}$
	$\nabla \times \boldsymbol{E} + \partial \boldsymbol{B}/\partial t = \boldsymbol{0}$	$\nabla \times \boldsymbol{E} + \frac{1}{c_0}\frac{\partial \boldsymbol{B}}{\partial t} = \boldsymbol{0}$
	$\nabla \cdot \boldsymbol{B} = 0$	$\nabla \cdot \boldsymbol{B} = 0$
Rate of radiation energy flow (Poynting vector):		
	$\boldsymbol{S} = \boldsymbol{E} \times \boldsymbol{H}$	$\boldsymbol{S} = c_0 \boldsymbol{E} \times \boldsymbol{H}$
Force on a moving charge Q with velocity $\boldsymbol{v}$ (Lorentz force):		
	$\boldsymbol{F} = Q(\boldsymbol{E} + \boldsymbol{v} \times \boldsymbol{B})$	$\boldsymbol{F} = Q(\boldsymbol{E} + \boldsymbol{v} \times \boldsymbol{B}/c_0)$

(1) $\boldsymbol{H}^{(ir)} = 4\pi\boldsymbol{H}$, $\boldsymbol{D}^{(ir)} = 4\pi\boldsymbol{D}$, ${\chi_e}^{(ir)} = \chi_e/4\pi$, $\chi^{(ir)} = \chi/4\pi$.

4 TABLES OF PHYSICAL QUANTITIES

The following tables contain the internationally recommended names and symbols for the physical quantities most likely to be used by chemists. Further quantities and symbols may be found in recommendations by IUPAP [4] and ISO [5].

Although authors are free to choose any symbols they wish for the quantities they discuss, provided that they define their notation and conform to the general rules indicated in Chapter 1, it is clearly an aid to scientific communication if we all generally follow a standard notation. The symbols below have been chosen to conform with current usage and to minimize conflict so far as possible. Small variations from the recommended symbols may often be desirable in particular situations, perhaps by adding or modifying subscripts or superscripts, or by the alternative use of upper or lower case. Within a limited subject area it may also be possible to simplify notation, for example by omitting qualifying subscripts or superscripts, without introducing ambiguity. The notation adopted should in any case always be defined. Major deviations from the recommended symbols should be particularly carefully defined.

The tables are arranged by subject. The five columns in each table give the name of the quantity, the recommended symbol(s), a brief definition by a mathematical equation, the symbol for the coherent SI unit (without multiple or submultiple prefixes, see Section 2.5, p. 14), and note references. When two or more symbols are recommended, commas are used to separate symbols that are equally acceptable, and symbols of second choice are put in parentheses. A semicolon is used to separate symbols of slightly different quantities. The names and symbols recommended here in the tables are in agreement with those recommended by IUPAP [4] and ISO [5.a-5.h].

The definitions are given primarily for identification purposes and are not necessarily complete. They should be regarded as useful relations rather than formal definitions. In some cases they imply a definition. For some of the quantities listed in this chapter, the definitions given in various IUPAC documents are collected in [21]. Useful definitions of physical quantities in physical organic chemistry can be found in [22] and those in polymer chemistry in [23]. For dimensionless quantities, a 1 is entered in the SI unit column. Further information is added in notes, and in text inserts between the tables, as appropriate. Other symbols used are defined within the same table (not necessarily in the order of appearance) and in the notes.

4.1 SPACE AND TIME

Name	*Symbols*	*Definition*	*SI unit*	*Notes*
Cartesian space coordinates	x; y; z		m	
cylindrical coordinates	ρ; ϑ; z		m, 1, m	
spherical polar coordinates	r; ϑ; φ		m, 1, 1	
generalized coordinates	q, q_i		(varies)	
position vector	$\boldsymbol{r}$	$\boldsymbol{r} = x\boldsymbol{e}_x + y\boldsymbol{e}_y + z\boldsymbol{e}_z$	m	1
length	l		m	
special symbols:				
height	h			
thickness	d, δ			
distance	d			
radius	r			
diameter	d			
path	$\boldsymbol{s}$			
length of arc	s			
area	A, A_s, S		m^2	1
volume	V, (v)		m^3	
plane angle	α, β, γ, ϑ, φ, ...	$\alpha = s/r$	rad, 1	2
solid angle	Ω, (ω)	$\Omega = A/r^2$	sr, 1	2
time, duration	t		s	
period	T	$T = t/N$	s	3
frequency	ν, f	$\nu = 1/T$	Hz, s^{-1}	
angular frequency	ω	$\omega = 2\pi\nu$	rad s^{-1}, s^{-1}	2, 4
relaxation time	τ, T	$\tau = \lvert \mathrm{d}t/\mathrm{d}\ln x \rvert$	s	
angular velocity	ω	$\omega = \mathrm{d}\varphi/\mathrm{d}t$	rad s^{-1}, s^{-1}	2, 5
velocity	$\boldsymbol{v}$, $\boldsymbol{u}$, $\boldsymbol{w}$, $\boldsymbol{c}$, $\dot{\boldsymbol{r}}$	$\boldsymbol{v} = \mathrm{d}\boldsymbol{r}/\mathrm{d}t$	m s^{-1}	
speed	v, u, w, c	$v = \lvert\boldsymbol{v}\rvert$	m s^{-1}	6
acceleration	$\boldsymbol{a}$	$\boldsymbol{a} = \mathrm{d}\boldsymbol{v}/\mathrm{d}t$	m s^{-2}	7

(1) $\boldsymbol{e}_x$, $\boldsymbol{e}_y$, and $\boldsymbol{e}_z$ are unit vectors in the space directions x, y, z. An infinitesimal area may be regarded as a vector $\boldsymbol{e}_\mathrm{n}\mathrm{d}A$, where $\boldsymbol{e}_\mathrm{n}$ is the unit vector normal to the plane.
(2) The units radian (rad) and steradian (sr) for plane angle and solid angle are derived. Since they are of dimension one (i.e. dimensionless), they may be included in expressions for derived SI units if appropriate, or omitted if clarity and meaning is not lost thereby.
(3) N is the number of identical (periodic) events during the time t.
(4) The unit Hz is not to be used for angular frequency.
(5) Angular velocity can be treated as a vector, $\boldsymbol{\omega}$, perpendicular to the plane of rotation defined by $\boldsymbol{\omega} = \boldsymbol{r} \times \boldsymbol{v} \,/\, r^2$.
(6) For the speeds of light and sound the symbol c is customary.
(7) For the modulus of the acceleration of gravity the symbol g is used. For the standard acceleration of gravity, the symbol g_n can be used (see inside front cover page).

4.2 CLASSICAL MECHANICS

Name	*Symbol*	*Definition*	*SI unit*	*Notes*
mass	m		kg	
reduced mass	μ	$\mu = m_1 m_2/(m_1 + m_2)$	kg	
density, mass density	ρ	$\rho = m/V$	kg m^{-3}	
relative density	d	$d = \rho/\rho^{\ominus}$	1	1
surface density	ρ_A, ρ_S	$\rho_A = m/A$	kg m^{-2}	
specific volume	v	$v = V/m = 1/\rho$	m^3 kg^{-1}	
momentum	$\boldsymbol{p}$	$\boldsymbol{p} = m\boldsymbol{v}$	kg m s^{-1}	
angular momentum	$\boldsymbol{L}$	$\boldsymbol{L} = \boldsymbol{r} \times \boldsymbol{p}$	J s	2
moment of inertia	I, J	$I = \sum_i m_i r_i^2$	kg m^2	3, 4
force	$\boldsymbol{F}$	$\boldsymbol{F} = \mathrm{d}\boldsymbol{p}/\mathrm{d}t = m\boldsymbol{a}$	N	
moment of force, torque	$\boldsymbol{M}, (\boldsymbol{T})$	$\boldsymbol{M} = \boldsymbol{r} \times \boldsymbol{F}$	N m	
energy	E		J	
potential energy	E_p, V, Φ	$E_\mathrm{p}(\boldsymbol{r}_1, \boldsymbol{r}_2) = -\int_{\boldsymbol{r}_1}^{\boldsymbol{r}_2} \boldsymbol{F} \cdot \mathrm{d}\boldsymbol{r}$	J	
kinetic energy	E_k, T, K	$E_\mathrm{k} = (1/2)mv^2$	J	
work	W, A, w	$W = \int \boldsymbol{F} \cdot \mathrm{d}\boldsymbol{r}$	J	
power	P	$P = \mathrm{d}W/\mathrm{d}t$	W	
generalized coordinate	q		(varies)	
generalized momentum	p		(varies)	
Hamilton function	H	$H(q,p) = T(q,p) + V(q)$	J	
pressure	$p, (P)$	$p = F/A$	Pa, N m^{-2}	
surface tension	γ, σ	$\gamma = \mathrm{d}W/\mathrm{d}A$	N m^{-1}, J m^{-2}	
weight	$G, (W, P)$	$G = mg$	N	
normal stress	σ	$\sigma = F/A$	Pa	3
shear stress	τ	$\tau = F/A$	Pa	3
linear strain	ε, e	$\varepsilon = \Delta l/l$	1	
modulus of elasticity, Young's modulus	E	$E = \sigma/\varepsilon$	Pa	3
shear strain	γ	$\gamma = \Delta x/d$	1	3, 5
shear modulus, Coulomb's modulus	G	$G = \tau/\gamma$	Pa	3
volume (bulk) strain	ϑ	$\vartheta = \Delta V/V_0$	1	3
bulk modulus, compression modulus	K	$K = -V_0\,(\mathrm{d}p/\mathrm{d}V)$	Pa	3
dynamic viscosity	$\eta, (\mu)$	$\tau_{xz} = \eta\,(\mathrm{d}v_x/\mathrm{d}z)$	Pa s	
fluidity	φ	$\varphi = 1/\eta$	m kg^{-1} s	
kinematic viscosity	ν	$\nu = \eta/\rho$	m^2 s^{-1}	
dynamic friction factor	$\mu, (f)$	$F_\mathrm{frict} = \mu F_\mathrm{norm}$	1	

(1) Usually $\rho^{\ominus} = \rho(\mathrm{H_2O}, 4\ ^\circ\mathrm{C})$.
(2) Other symbols are customary in atomic and molecular spectroscopy (see Section 4.6, p. 29).
(3) In general these can be tensor quantities.
(4) In general $\boldsymbol{I}$ is a tensor quantity: $I_{\alpha\alpha} = \sum_i m_i\left(\beta_i^2 + \gamma_i^2\right)$, and $I_{\alpha\beta} = -\sum_i m_i \alpha_i \beta_i$ if $\alpha \neq \beta$, where α, β, γ is a permutation of x, y, z.
(5) d is the distance between the layers displaced by Δx.

4.3 ELECTRICITY AND MAGNETISM

Name	*Symbol*	*Definition*	*SI unit*	*Notes*
electric current	I, i		A	
electric current density	$\boldsymbol{j}, \boldsymbol{J}$	$I = \int \boldsymbol{j} \cdot \boldsymbol{e}_\mathrm{n}\, \mathrm{d}A$	A m^{-2}	1
electric charge	Q	$Q = \int I\, \mathrm{d}t$	C	
charge density	ρ	$\rho = Q/V$	C m^{-3}	
electric potential	V, ϕ	$V = \mathrm{d}W/\mathrm{d}Q$	V, J C^{-1}	
electric potential difference, electric tension	$U, \Delta V, \Delta\phi$	$U = V_2 - V_1$	V	
electric field strength	$\boldsymbol{E}$	$\boldsymbol{E} = \boldsymbol{F}/Q = -\boldsymbol{\nabla} V$	V m^{-1}	
electric displacement	$\boldsymbol{D}$	$\boldsymbol{\nabla} \cdot \boldsymbol{D} = \rho$	C m^{-2}	
capacitance	C	$C = Q/U$	F, C V^{-1}	
permittivity	ε	$\boldsymbol{D} = \varepsilon \boldsymbol{E}$	F m^{-1}	2
electric constant	ε_0	$\varepsilon_0 = {\mu_0}^{-1} {c_0}^{-2}$	F m^{-1}	3
dielectric polarization	$\boldsymbol{P}$	$\boldsymbol{P} = \boldsymbol{D} - \varepsilon_0 \boldsymbol{E}$	C m^{-2}	
electric susceptibility	χ_e	$\chi_\mathrm{e} = (\varepsilon - \varepsilon_0)/\varepsilon_0$	1	
electric dipole moment	$\boldsymbol{p}, \boldsymbol{\mu}$	$\boldsymbol{p} = \sum_i Q_i \boldsymbol{r}_i$	C m	4
magnetic flux density, magnetic induction	$\boldsymbol{B}$	$\boldsymbol{F} = Q \boldsymbol{v} \times \boldsymbol{B}$	T	
magnetic flux	Φ	$\Phi = \int \boldsymbol{B} \cdot \boldsymbol{e}_\mathrm{n}\, \mathrm{d}A$	Wb	1
magnetic field strength	$\boldsymbol{H}$	$\nabla \times \boldsymbol{H} = \boldsymbol{j}$	A m^{-1}	
permeability	μ	$\boldsymbol{B} = \mu \boldsymbol{H}$	N A^{-2}, H m^{-1}	2
magnetic constant	μ_0	$\mu_0 \approx 4\pi \times 10^{-7}$ H m^{-1}	H m^{-1}	
relative permeability	μ_r	$\mu_\mathrm{r} = \mu/\mu_0$	1	
magnetization	$\boldsymbol{M}$	$\boldsymbol{M} = \boldsymbol{B}/\mu_0 - \boldsymbol{H}$	A m^{-1}	
magnetizability of a molecule	ξ	$\boldsymbol{m} = \xi \boldsymbol{B}$	J T^{-2}	
magnetic susceptibility	χ	$\chi = \mu_\mathrm{r} - 1$	1	
molar magnetic susceptibility	χ_m	$\chi_\mathrm{m} = V_\mathrm{m} \chi$	m^3 mol^{-1}	5
magnetic dipole moment	$\boldsymbol{m}, \boldsymbol{\mu}$	$E_\mathrm{p} = -\boldsymbol{m} \cdot \boldsymbol{B}$	A m^2, J T^{-1}	6
electric resistance	R	$R = U/I$	Ω	7
conductance	G	$G = 1/R$	S	
conductivity	κ	$\boldsymbol{j} = \kappa \boldsymbol{E}$	S m^{-1}	2
self-inductance	L	$E = -L(\mathrm{d}I/\mathrm{d}t)$	H, V s A^{-1}	
magnetic vector potential	$\boldsymbol{A}$	$\boldsymbol{B} = \boldsymbol{\nabla} \times \boldsymbol{A}$	Wb m^{-1}	
Poynting vector	$\boldsymbol{S}$	$\boldsymbol{S} = \boldsymbol{E} \times \boldsymbol{H}$	W m^{-2}	8

(1) $\boldsymbol{e}_\mathrm{n}\mathrm{d}A$ is a vector element of area.
(2) This quantity is a second-rank tensor in anisotropic materials.
(3) c_0 is the speed of light in vacuum.
(4) When a dipole is composed of two point charges Q and $-Q$ separated by a distance r, the direction of the dipole vector is taken to be from the negative to the positive charge. The opposite convention is sometimes used, but is to be discouraged. The dipole moment of an ion depends on the choice of the origin. An alternative definition is $E_\mathrm{p} = -\boldsymbol{p} \cdot \boldsymbol{E}$.
(5) See also footnote 2 on p. 4.
(6) E_p is the potential energy of the dipole.
(7) In a material with non zero reactance $X = (U/I) \sin\delta$ and loss angle δ, $R = (U/I) \cos\delta$ and $G = R/(R^2 + X^2)$.
(8) This quantity is also called the Poynting-Umov vector.

4.4 QUANTUM MECHANICS AND QUANTUM CHEMISTRY

Angular momentum operators are treated in Section 4.6.3, p. 33.

Name	Symbol	Definition	SI unit	Notes
momentum operator	$\hat{\boldsymbol{p}}$	$\hat{\boldsymbol{p}} = -\mathrm{i}\hbar\,\boldsymbol{\nabla}$	$\mathrm{J\,s\,m^{-1}}$	1
kinetic energy operator	$\hat{T}$	$\hat{T} = -(\hbar^2/2m)\boldsymbol{\nabla}^2$	J	1
Hamiltonian operator, Hamiltonian	$\hat{H}$	$\hat{H} = \hat{T} + \hat{V}$	J	1
wavefunction, state function	Ψ, ψ, ϕ	$\hat{H}\psi = E\psi$	$(\mathrm{m}^{-3/2})$	2
hydrogen-like wavefunction	$\psi_{nlm}(r,\theta,\phi)$	$\psi_{nlm} = R_{nl}(r)Y_{lm}(\theta,\phi)$	$(\mathrm{m}^{-3/2})$	2
spherical harmonic function	$Y_{lm}(\theta,\phi)$	$Y_{lm} = N_{l\lvert m\rvert}P_l^{\lvert m\rvert}(\cos\theta)\mathrm{e}^{\mathrm{i}m\phi}$	1	3
probability density	P	$P = \psi^*\psi$	(m^{-3})	2, 4
probability current density	$\boldsymbol{S}$	$\boldsymbol{S} = -(\mathrm{i}\hbar/2m)\times(\psi^*\boldsymbol{\nabla}\psi - \psi\boldsymbol{\nabla}\psi^*)$	$(\mathrm{m^{-2}\,s^{-1}})$	2
integration element	$\mathrm{d}\tau$	$\mathrm{d}\tau = \mathrm{d}x\ \mathrm{d}y\ \mathrm{d}z$	(varies)	
matrix element of operator $\hat{A}$	A_{ij}, $\left\langle i \middle\lvert \hat{A} \middle\rvert j \right\rangle$	$A_{ij} = \int \psi_i{}^*\hat{A}\psi_j \mathrm{d}\tau$	(varies)	5
Hermitian conjugate of operator $\hat{A}$	$\hat{A}^\dagger$	$\left(A^\dagger\right)_{ij} = (A_{ji})^*$	(varies)	5
commutator of $\hat{A}$ and $\hat{B}$	$[\hat{A},\hat{B}]$, $[\hat{A},\hat{B}]_-$	$[\hat{A},\hat{B}] = \hat{A}\hat{B} - \hat{B}\hat{A}$	(varies)	6
anticommutator of $\hat{A}$ and $\hat{B}$	$[\hat{A},\hat{B}]_+$	$[\hat{A},\hat{B}]_+ = \hat{A}\hat{B} + \hat{B}\hat{A}$	(varies)	6
spin wavefunction	α; β		1	7

(1) The circumflex (or "hat"), $\hat{}$, serves to distinguish an operator from an algebraic quantity. The definition of the momentum and kinetic energy operators applies to the coordinate representation, where $\boldsymbol{\nabla}$ denotes the nabla operator (see Section 1.8, p. 9).
(2) For the normalized wavefunction of a single particle in three-dimensional space the appropriate SI unit is given in parentheses.
(3) $P_l^{\lvert m\rvert}$ denotes the associated Legendre function of degree l and order $\lvert m\rvert$. $N_{l\lvert m\rvert}$ is a normalization factor.
(4) ψ^* is the complex conjugate of ψ.
(5) The unit is the same as for the physical quantity A that the operator represents. The diagonal element A_{ii} is also called the *expectation value*, $\langle A\rangle$ or $\bar{A}$, of the operator $\hat{A}$ in the state ψ_i.
(6) The unit is the same as for the product of the physical quantities A and B.
(7) The spin wavefunctions of a single electron, α and β, are defined by the matrix elements of the z component of the spin angular momentum, $\hat{s}_z$, by the relations $\langle\alpha\,\lvert\,\hat{s}_z\rvert\,\alpha\rangle = +(1/2)$, $\langle\beta\,\lvert\,\hat{s}_z\rvert\,\beta\rangle = -(1/2)$, $\langle\beta\,\lvert\,\hat{s}_z\rvert\,\alpha\rangle = \langle\alpha\,\lvert\,\hat{s}_z\rvert\,\beta\rangle = 0$ in units of $\hbar$.

4.4.1 Symbols commonly used in quantum chemistry

A list of acronyms used in theoretical chemistry has been published by IUPAC [24]. Results in quantum chemistry are typically expressed in atomic units. In the remaining tables of this section all lengths, energies, masses, charges and angular momenta are expressed as dimensionless ratios to the corresponding atomic units, a_0, E_h, m_e, e and $\hbar$ respectively. Thus all quantities become dimensionless, and the SI unit column is therefore omitted.

Name	*Symbol*	*Definition*	*Notes*
(atomic) basis function	$\chi_r(\mu)$		8
orbital	ϕ_i	$\phi_i(\mu) = \sum_r \chi_r(\mu) c_{ri}$	8
spin orbital	$\phi_i(\mu)\,\alpha(\mu)$; $\phi_i(\mu)\,\beta(\mu)$		9
configuration function	Φ	$\Phi = \sum_I c_I \Phi_I$	10
general wavefunction	Ψ	$\Psi = (n!)^{-1/2} \lVert\phi_i(\mu)\rVert$	10
core Hamiltonian of a single electron	$\hat{H}_\mu{}^{\text{core}}$	$\hat{H}_\mu = -(1/2)\nabla_\mu^2 - \sum_A Z_A/r_{\mu A}$	11
overlap matrix element	S_{rs}	$S_{rs} = \int \chi_r{}^* \chi_s \mathrm{d}\tau, \quad \sum_{r,s} c_{ri}{}^* S_{rs} c_{sj} = \delta_{ij}$	8
density matrix element	P_{rs}	$P_{rs} = 2\sum_i^{\text{occ}} c_{ri}{}^* c_{si}$	8, 12
one-electron integrals	H_{rs}	$H_{rs} = \int \chi_r{}^*(1) \hat{H}_1{}^{\text{core}} \chi_s(1)\mathrm{d}\tau_1$	
two-electron integrals	$(rs\vert tu)$	$(rs\vert tu) = \iint \chi_r{}^*(1)\,\chi_s(1)\,\frac{1}{r_{12}}\chi_t{}^*(2)\,\chi_u(2)\,\mathrm{d}\tau_1\mathrm{d}\tau_2$	13
matrix element of the Fock operator	F_{rs}	$F_{rs} = H_{rs} + \sum_t \sum_u P_{tu}\left[(rs\vert tu) - (1/2)\,(ru\vert ts)\right]$	14
orbital energy	ε_i	$\varepsilon_i = H_{ii} + \sum_j (2J_{ij} - K_{ij})$	12, 15
configuration state energy	E_I	$E_I = \sum_{i \in I} (\varepsilon_i + H_{ii})$	12, 15
energy parameter	x	$x = (\varepsilon - \alpha)/\beta$	16

(8) The indices i and j label the molecular orbitals, and either μ or the numerals 1 and 2 label the electron coordinates. The indices r and s label the basis functions. The expansion coefficients c_{ri}, c_{sj}, etc. are complex numbers.
(9) α and β are the spin functions; the spin orbitals $\phi_i\alpha$ and $\phi_i\beta$ are also denoted ϕ_i and $\overline{\phi}_i$.
(10) The symbol $\lVert\phi_i(\mu)\rVert$ denotes an anti-symmetrized product of n occupied molecular spin orbitals $\phi_i\alpha$ and $\phi_i\beta$, n is the number of electrons. The index I describes the n-tuple of indices of occupied spin orbitals. The configuration function is the wavefunction of the configuration state.
(11) Z_A is the proton number (charge number) of nucleus A, and $r_{\mu A}$ is the distance of electron μ from nucleus A.
(12) The definition given here applies to closed-shell systems.
(13) The two-electron integrals are written in various shorthand notations: In $J_{ij} = \langle ij\vert ij\rangle$ the first and third indices refer to the index of electron 1 and the second and fourth indices to electron 2. In $J_{ij} = (i^*i\vert j^*j)$, the first two indices refer to electron 1 and the second two indices to electron 2. Usually the functions are real and the stars are omitted.
(14) For closed-shell systems, optimized orbitals are self-consistent solutions of the non-linear Hartree–Fock–Roothaan equations $\sum_s (F_{rs} - \varepsilon_i S_{rs})\,c_{si} = 0$.
(15) H_{ii} is the expectation value of the core Hamiltonian in orbital i. $J_{ij} = (i^*i\vert j^*j) = \langle ij\vert ij\rangle$ is the *Coulomb integral*, and $K_{ij} = (i^*j\vert j^*i) = \langle ij\vert ji\rangle$ is the *exchange integral*. The definition applies to optimized orbitals.
(16) This quantity is used in the *Hückel molecular orbital theory* (HMO); $\alpha = H_{rr}$ and $\beta = H_{rs}$, where the atomic basis functions χ_r belong to one, χ_s to the other partner of a pair of bonded atoms. In HMO theory, all H_{rr} are equal, and $H_{rs} = 0$ and $S_{rs} = 0$ for non-bonded pairs of atoms.

4.5 ATOMS AND MOLECULES

Name	*Symbol*	*Definition*	*SI unit*	*Notes*
nucleon number, mass number	A		1	
proton number, atomic number	Z		1	
neutron number	N	$N = A - Z$	1	
mass of atom, atomic mass	m_a, m		kg	
ionization energy	E_i, I		J	1
electron affinity	E_{ea}, A		J	1
electronegativity	χ	$\chi = (1/2)(E_i + E_{ea})$	J	2
dissociation energy	E_d, D		J	
from the ground state	D_0		J	3
from the potential minimum	D_e		J	3
principal quantum number (hydrogen-like atom)	n	$E = hcZ^2R_\infty/n^2$	1	4
magnetic dipole moment of a molecule	$\boldsymbol{m}, \boldsymbol{\mu}$	$E_p = -\boldsymbol{m} \cdot \boldsymbol{B}$	$J\ T^{-1}$	5
magnetizability of a molecule	ξ	$\boldsymbol{m} = \xi \boldsymbol{B}$	$J\ T^{-2}$	
Bohr magneton	μ_B	$\mu_B = e\hbar/2m_e$	$J\ T^{-1}$	
nuclear magneton	μ_N	$\mu_N = e\hbar/2m_p = (m_e/m_p)\,\mu_B$	$J\ T^{-1}$	
gyromagnetic ratio, (magnetogyric ratio)	γ	$\gamma_e = -g_e\mu_B/\hbar$	$s^{-1}\ T^{-1}$	6
g-factor	g, g_e	$g_e = -\gamma_e(2m_e/e)$	1	7
nuclear g-factor	g_N	$g_N = \gamma_N(2m_p/e)$	1	7
Larmor angular frequency	$\boldsymbol{\omega}_L$	$\boldsymbol{\omega}_L = -\gamma \boldsymbol{B}$	s^{-1}	8
electric dipole moment of a molecule	$\boldsymbol{p}, \boldsymbol{\mu}$	$E_p = -\boldsymbol{p} \cdot \boldsymbol{E}$	C m	5
quadrupole moment	$\boldsymbol{Q}$	$Q_{\alpha\beta} = \int r_\alpha r_\beta \rho \,\mathrm{d}V$	$C\ m^2$	9
quadrupole moment	eQ		$C\ m^2$	10

(1) The ionization energy is frequently called the ionization potential (I_p). The electron affinity is the energy released in attaching an electron.
(2) There are several ways of defining this quantity [25], the one given in the table has a clear physical meaning of energy and is due to R. S. Mulliken.
(3) The symbols D_0 and D_e are used for dissociation energies of diatomic and polyatomic molecules.
(4) For an electron in the central Coulomb field of an infinitely heavy nucleus of atomic number Z. R_∞ is the Rydberg constant, $R_\infty = E_h/2hc_0$.
(5) E_p is the potential energy of the dipole. For ions, the electric dipole moment depends on the choice of the origin.
(6) The gyromagnetic ratio for a nucleus is $\gamma_N = g_N\mu_N/\hbar$.
(7) For historical reasons, $g_e > 0$. e is the (positive) elementary charge, therefore $\gamma_e < 0$. For nuclei, γ_N and g_N have the same sign. A different sign convention for the electronic g-factor is discussed in [26].
(8) This is a vector quantity with magnitude ω_L and is sometimes called Larmor circular frequency. The quantity $\nu_L = \omega_L/2\pi$ (in Hz) is called Larmor frequency.
(9) The quadrupole moment of a molecule may be represented by the tensor $\boldsymbol{Q}$, defined by an integral over the charge density ρ. $r_{\alpha,\beta}$ are Cartesian space coordinates.
(10) The nuclear quadrupole moment Q of a nucleus has the dimension of an area and e is the elementary charge. Here, eQ is used as a common symbol for 'nuclear quadrupole moment'.

Name	*Symbol*	*Definition*	*SI unit*	*Notes*
electric field gradient tensor	$\boldsymbol{q}$	$q_{\alpha\beta} = -\partial^2 V/\partial\alpha\partial\beta$	$V\ m^{-2}$	
quadrupole interaction energy tensor	$\boldsymbol{\chi}$	$\chi_{\alpha\beta} = eQq_{\alpha\beta}$	J	11
electric polarizability of a molecule	$\boldsymbol{\alpha}$	$\alpha_{ab} = \partial p_a/\partial E_b$	$C^2\ m^2\ J^{-1}$	12
(radio)activity	A	$A = -\mathrm{d}N_\mathrm{B}/\mathrm{d}t$	Bq	13
decay (rate) constant	λ, k	$\lambda = A/N_\mathrm{B}$	s^{-1}	13

(11) The nuclear quadrupole interaction energy tensor $\boldsymbol{\chi}$ is usually quoted in MHz, corresponding to the value of eQq/h, although the h is usually omitted in the notation.
(12) The polarizability $\boldsymbol{\alpha}$ (and the hyper-polarizabilities $\boldsymbol{\beta}, \boldsymbol{\gamma}, \cdots$) are the coefficients in the expansion of the dipole moment $\boldsymbol{p}$ in powers of the electric field strength $\boldsymbol{E}$ (see Section 2.3 and 2.5 in [20]).
(13) N_B is the number of decaying entities B ($1\ \mathrm{Bq} = 1\ \mathrm{s}^{-1}$). Further quantities related to the decay constant such as lifetime τ and half life $t_{1/2}$ can be found in Section 4.12, p. 54.

4.6 SPECTROSCOPY

This section is based on the recommendations of the ICSU Joint Commission for Spectroscopy [27,28] and current practice in the field which is well represented in the books by Herzberg [29–31], in the 'Handbook of High-Resolution Spectroscopy' [32] and by IUPAC recommendations [33–44].

4.6.1 Optical Spectroscopy

Name	*Symbol*	*Definition*	*SI unit*	*Notes*
total term	T	$T = E_{\text{tot}}/hc$	m^{-1}	1, 2
transition wavenumber, repetency	$\tilde{\nu}$	$\tilde{\nu} = T' - T''$	m^{-1}	1
transition frequency	ν	$\nu = (E' - E'')/h$	Hz	1
vibrational quantum numbers	$v_r; l_t$		1	3
vibrational fundamental wavenumber	$\tilde{\nu}_r, \tilde{\nu}_r^0, (\nu_r)$	$\tilde{\nu}_r = T_{(v_r=1)} - T_{(v_r=0)}$	m^{-1}	3
Coriolis ζ-constant	ζ_{rs}^{α}		1	4
degeneracy, statistical weight	g, d, β		1	5
harmonic vibration wavenumber	$\omega_\text{e}; \omega_r$		m^{-1}	3
vibrational anharmonicity constant	$\omega_\text{e}x_\text{e}; x_{rs}; g_{tt'}$		m^{-1}	3
principal moments of inertia	$I_A; I_B; I_C$	$I_A \leqslant I_B \leqslant I_C$	kg m^2	
rotational constant	$A; B; C$	$A = h/8\pi^2 I_A$	Hz	1,2

(1) In spectroscopy the unit cm^{-1} is almost always used for the quantity wavenumber, thus term value and wavenumber always refer to the reciprocal wavelength of the equivalent radiation in vacuum, $\tilde{\nu} = 1/\lambda$. Superscript $'$ signifies the upper state in a transition and superscript $''$ the lower state, for example also v_r' and v_r''. The symbol c in the definition E/hc refers to the speed of light in vacuum. The use of the word "wavenumber" in place of the unit cm^{-1} must be avoided.
(2) Term values and rotational constants are sometimes defined as wavenumber (e.g. $T = E/hc$), and sometimes as frequency (e.g. $T = E/h$). When the symbol is otherwise the same, it is convenient to distinguish wavenumber quantities with a tilde (e.g. $\tilde{\nu}, \tilde{T}, \tilde{A}, \tilde{B}, \tilde{C}$, etc.) and the conversion $\tilde{\nu} = \nu/c$, $\tilde{B} = B/c$, etc. Specific term values are for electronic ($\tilde{T}_\text{e} = E_\text{e}/hc$), vibrational ($\tilde{G} = E_{\text{vib}}/hc$) and rotational ($\tilde{F} = E_{\text{rot}}/hc$) energies.
(3) For a diatomic molecule: $G(v) = \omega_\text{e}\left(v + \frac{1}{2}\right) - \omega_\text{e}x_\text{e}\left(v + \frac{1}{2}\right)^2 + \cdots$. For a polyatomic molecule the $3N-6$ vibrational modes ($3N-5$ if linear) are labeled by the indices $r, s, t, \cdots$, or $i, j, k, \cdots$. The index r is usually assigned to be increasing with descending wavenumber, symmetry species by irreducible representation labels starting with the totally symmetric species. The index t is kept for degenerate modes. The vibrational term formula is

$$G(v) = \sum_r \omega_r (v_r + d_r/2) + \sum_{r \leqslant s} x_{rs} (v_r + d_r/2)(v_s + d_s/2) + \sum_{t \leqslant t'} g_{tt'} l_t l_{t'} + \cdots$$

Another common term formula is defined with respect to the vibrational ground state, see Section 2.6 in [20].
(4) Frequently the Coriolis coupling constants ${\xi_\alpha}^{v'v}$ with units of cm^{-1} are used as effective Hamiltonian constants ($\alpha = x, y, z$). For two fundamental vibrations with harmonic wavenumbers ω_r and ω_s these are connected with ${\zeta_{rs}}^\alpha$ by the equation ($v_r = 1$ and $v_s = 1$)

$${\xi_\alpha}^{v'v} = \tilde{B}_\alpha\, {\zeta_{rs}}^\alpha \left[\sqrt{\omega_s/\omega_r} + \sqrt{\omega_r/\omega_s}\right]$$

where $\tilde{B}_\alpha$ is the α rotational constant. A similar equation applies with B_α if ${\xi_\alpha}^{v'v}$ is defined as a quantity with frequency units.

Name	*Symbol*	*Definition*	*SI unit*	*Notes*
asymmetry parameter	κ	$\kappa = \dfrac{2B - A - C}{A - C}$	1	
centrifugal distortion constants,				
S reduction	$D_J; D_{JK}; D_K; d_1; d_2$		m^{-1}	6
A reduction	$\Delta_J; \Delta_{JK}; \Delta_K; \delta_J; \delta_K$		m^{-1}	6
spin-orbit coupling constant	A	$T_{so} = A < \widehat{\boldsymbol{L}} \cdot \widehat{\boldsymbol{S}} >$	m^{-1}	1, 7
transition dipole moment of a molecule	$\boldsymbol{M}_{ij}, \boldsymbol{R}_{ij}$	$\boldsymbol{M}_{ij} = \int \psi_i^* \boldsymbol{p}\, \psi_j\, d\tau$	C m	8
interatomic distances,				9, 10
equilibrium	r_e		m	
ground state	r_0		m	
vibrational coordinates,				9
internal	R_i, r_i, θ_j, etc.		(varies)	
normal				
mass adjusted	Q_r		$kg^{1/2}$ m	
dimensionless	q_r		1	

(5) d is usually used for vibrational, and β for nuclear spin degeneracy.
(6) S and A stand for the symmetric and asymmetric reductions of the rotational Hamiltonian respectively; see [45] for more details on the various possible representations of the centrifugal distortion constants (see also [32]).
(7) $\widehat{\boldsymbol{L}}$ and $\widehat{\boldsymbol{S}}$ are electron orbital and electron spin operators, respectively.
(8) $\boldsymbol{p}$ is the electric or magnetic dipole moment of a molecule. The given unit holds for the electric dipole moment.
(9) Interatomic (internuclear) distances and vibrational displacements are sometimes expressed in the non-SI unit ångström, Å, where 1 Å $= 10^{-10}$ m $= 0.1$ nm $= 100$ pm. 100 pm should be preferred over Å. The symbol S_j is often used for symmetry coordinates.
(10) Different ways of representing interatomic distances exist and are distinguished by subscripts (see [20]). Only the equilibrium distance r_e is isotopically invariant, to a good approximation.

4.6.2 Electron paramagnetic resonance (EPR), electron spin resonance (ESR), and nuclear magnetic resonance (NMR)

Name	*Symbol*	*Definition*	*SI unit*	*Notes*
gyromagnetic ratio	γ	$\gamma = \mu/\hbar\sqrt{S(S+1)}$	$s^{-1}\ T^{-1}$	11
g-factor	g	$h\nu = g\mu_B B$	1	12
hyperfine coupling constant in liquids	a, A	$\widehat{H}_{hfs}/h = a\,\widehat{\boldsymbol{S}}\cdot\widehat{\boldsymbol{I}}$	Hz	13
magnetic flux densities	$\boldsymbol{B}_0, \boldsymbol{B}_1, \boldsymbol{B}_2$		T	14
spin-rotation coupling constant of nucleus A	C_A		Hz	15
dipolar coupling constant between nuclei A, B	D_{AB}	$D_{AB} = \frac{\mu_0\hbar}{8\pi^2}\frac{\gamma_A\gamma_B}{r_{AB}^{\ 3}}$	Hz	15
frequency variables of a two-dimensional spectrum	$F_1, F_2; f_1, f_2$		Hz	
nuclear spin-spin coupling through n bonds	${}^nJ$		Hz	16
reduced nuclear spin-spin coupling constant	K_{AB}	$K_{AB} = \frac{J_{AB}}{h}\frac{2\pi}{\gamma_A}\frac{2\pi}{\gamma_B}$	$T^2\ J^{-1}$, $N\ A^{-2}\ m^{-3}$	
equilibrium macroscopic magnetization per volume	$\boldsymbol{M}_0$		$J\ T^{-1}\ m^{-3}$	17
nuclear quadrupole moment	eQ		$C\ m^2$	18
electric field gradient	$\boldsymbol{q}$	$q_{\alpha\beta} = -\partial^2 V/\partial\alpha\partial\beta$	$V\ m^{-2}$	19

(11) The magnitude of γ is obtained from the magnitudes of the magnetic dipole moment μ (of the electron for S or the nucleus for I) and the angular momentum (in ESR: electron spin $\boldsymbol{S}$ for a Σ-state, $L = 0$; in NMR: nuclear spin $\boldsymbol{I}$). μ is the electric dipole moment of the electrons in ESR, and of the nuclei in NMR.

(12) This gives an effective g-factor for a single spin ($S = 1/2$) in an external static field. A convention for the g-factor has been proposed [26], see Section 4.5, note 7, p. 27.

(13) $\widehat{H}_{hfs}$ is the hyperfine coupling Hamiltonian. $\widehat{\boldsymbol{S}}$ is the electron spin operator with quantum number S (see Section 4.6.1, p. 33). The coupling constants a are usually quoted in MHz, but they are sometimes quoted in magnetic induction units (G or T) with the conversion factor $g_e\mu_B/h \approx 28.025$ GHz T^{-1} $\left(= 2.8025 \text{ MHz } G^{-1}\right)$, where g_e is the g-factor for a free electron. If in liquids the hyperfine coupling is isotropic, the coupling constant is a scalar a. In solids the coupling is anisotropic, and the coupling constant is a tensor (similarly for the g-factor and to the analogous NMR parameters). A convention for the g-factor has been proposed, in which the sign of g is positive when the dipole moment is parallel to its angular momentum and negative, when it is antiparallel. Such a choice would require the g-factors for the electron orbital and spin angular momenta to be negative [26] (see Section 4.5, note 7, p. 27).

(14) $\boldsymbol{B}_1$ denotes the observing and $\boldsymbol{B}_2$ the irradiating radiofrequency flux densities; these are associated with frequency ν_i and with nutation angular frequency Ω_i with $i = 1, 2$ (around $\boldsymbol{B}_i$, respectively). They are defined through $\Omega_i = -\gamma\boldsymbol{B}_i$ (see Section 4.3, note 4, p. 24). $\boldsymbol{B}_0$ denotes the static magnetic flux density of a NMR spectrometer.

(15) The units of interaction strengths are Hz. In the context of relaxation the interaction strengths should be converted to angular frequency units (rad s^{-1}, but commonly denoted s^{-1}, see note 20).

(16) Parentheses may be used to indicate the species of the nuclei coupled, e.g. $J(^{13}C, ^1H)$ or, additionally, the coupling path, e.g. J(POCF). Where no ambiguity arises, the elements involved can be, alternatively, given as subscripts, e.g. J_{CH}. The nucleus of higher mass should be given first. The same applies to the reduced coupling constant.

(17) This is for a spin system in the presence of a magnetic flux density $\boldsymbol{B}_0$.

(18) See Section 4.5, notes 9 and 10, p. 27.

(19) The symbol $\boldsymbol{q}$ is recommended by IUPAC for the field gradient tensor, with the units of $V\ m^{-2}$ (see Section 4.5, p. 27). With $\boldsymbol{q}$ defined in this way, the quadrupole coupling constant is $\chi = eq_{zz}Q/h$.

Name	*Symbol*	*Definition*	*SI unit*	*Notes*
nuclear quadrupole coupling constant	χ	$\chi = eq_{zz}Q/h$	Hz	15
spin-lattice relaxation time			s	20, 21
(longitudinal) for nucleus A	T_1^{A}			
(transverse) for nucleus A	T_2^{A}			
gyromagnetic ratio	γ	$\gamma = \mu/\hbar\sqrt{I(I+1)}$	$\mathrm{s^{-1}\ T^{-1}}$	11
chemical shift for the nucleus A	δ_{A}	$\delta_{\mathrm{A}} = (\nu_{\mathrm{A}} - \nu_{\mathrm{ref}})/\nu_{\mathrm{ref}}$	1	22
nuclear magneton	μ_{N}	$\mu_{\mathrm{N}} = (m_{\mathrm{e}}/m_{\mathrm{p}})\mu_{\mathrm{B}}$	$\mathrm{J\ T^{-1}}$	23
standardized resonance frequency for nucleus A	Ξ_{A}		Hz	24
shielding constant	σ_{A}	$B_{\mathrm{A}} = (1 - \sigma_{\mathrm{A}})B_0$	1	25
correlation time	τ_{c}		s	20

(19) (continued)
It is common in NMR to denote the field gradient tensor as $e\boldsymbol{q}$ and the quadrupole coupling constant as $\chi = e^2q_{zz}Q/h$. $\boldsymbol{q}$ has principal components q_{XX}, q_{YY}, q_{ZZ}.

(20) The relaxation times and the correlation times are normally given in the units of s. Strictly speaking, this refers to s rad^{-1}.

(21) The spin-lattice relaxation time of nucleus A in the frame of reference rotating with $\boldsymbol{B}_1$ is denoted $T_{1\rho}{}^{\mathrm{A}}$.

(22) Chemical shift (of the resonance) for the nucleus of element A (positive when the sample resonates to high frequency of the reference); for conventions, see [46]. ν_{A} and ν_{ref} are resonance frequencies.

(23) m_{e} and m_{p} are the electron and proton mass, respectively. μ_{B} is the Bohr magneton (see Section 4.5, p. 27).

(24) Resonance frequency for the nucleus of element A in a magnetic field such that the protons in tetramethylsilane (TMS) resonate exactly at 100 MHz.

(25) The symbols σ_{A} (and related terms of the shielding tensor and its components) should refer to shielding on an absolute scale (for theoretical work). For shielding relative to a reference, symbols such as $\sigma_{\mathrm{A}} - \sigma_{\mathrm{ref}}$ should be used. B_{A} is the corresponding effective magnetic flux density (see Section 4.3, p. 24).

4.6.3 Symbols for angular momentum operators and quantum numbers

In the following table, all of the operator symbols denote the dimensionless ratio *angular momentum* divided by $\hbar$. The column heading "*Z-axis*" denotes the space-fixed component, and the heading "*z-axis*" denotes the molecule-fixed component along the symmetry axis (linear or symmetric top molecules), or the axis of quantization.

Angular momentum[1]	*Operator symbol*	*Quantum number symbol* *Total*	*Z-axis*	*z-axis*	*Notes*
electron orbital	$\widehat{\boldsymbol{L}}$	L	M_L	Λ	2
one electron only	$\widehat{\boldsymbol{l}}$	l	m_l	λ	2
electron spin	$\widehat{\boldsymbol{S}}$	S	M_S	Σ	
one electron only	$\widehat{\boldsymbol{s}}$	s	m_s	σ	
electron orbital plus spin	$\widehat{\boldsymbol{L}}+\widehat{\boldsymbol{S}}$			$\Omega=\Lambda+\Sigma$	2
nuclear orbital (rotational)	$\widehat{\boldsymbol{R}}$	R		K_R, k_R	
nuclear spin	$\widehat{\boldsymbol{I}}$	I	M_I		
internal vibrational					
spherical top	$\widehat{\boldsymbol{l}}$	$l\,(l\zeta)$		K_l	3
other	$\widehat{\boldsymbol{j}}, \widehat{\boldsymbol{\pi}}$			$l\,(l\zeta)$	2, 3
sum of $R+L\,(+j)$	$\widehat{\boldsymbol{N}}$	N		K, k	2
sum of $N+S$	$\widehat{\boldsymbol{J}}$	J	M_J	K, k	2, 4
sum of $J+I$	$\widehat{\boldsymbol{F}}$	F	M_F		

(1) In all cases the vector operator and its components are related to the quantum numbers by eigenvalue equations analogous to:

$$\widehat{J}^2\psi = J\,(J+1)\,\psi,\quad \widehat{J}_Z\psi = M_J\psi,\quad \text{and}\quad \widehat{J}_z\,\psi = K\psi,$$

The component quantum numbers M_J and K take integral or half-odd values in the range $-J \leqslant M_J \leqslant +J$, $-J \leqslant K \leqslant +J$. l is often called the azimuthal quantum number and m_l the magnetic quantum number.
(2) Some authors, notably Herzberg [29–31], treat the component quantum numbers Λ, Ω, l and K as taking positive or zero values only, so that each non-zero value of the quantum number labels two wavefunctions with opposite signs for the appropriate angular momentum component. When this is done, lower case k is commonly regarded as a signed quantum number, related to K by $K = |k|$. However, in theoretical discussions all component quantum numbers are usually treated as signed, taking both positive and negative values.
(3) There is no uniform convention for denoting the internal vibrational angular momentum. For symmetric top and linear molecules the component of j in the symmetry axis is always denoted by the quantum number l, where l takes values in the range $-v \leqslant l \leqslant +v$ in steps of 2. The corresponding component of angular momentum is actually $l\zeta$, rather than l, where ζ is the Coriolis ζ-constant (see note 4, p. 29).
(4) Asymmetric top rotational states are labeled by the value of J (or N if $S \neq 0$), with subscripts K_a, K_c, (for example $J_{K_a,K_c} = 5_{2,3}$) where the latter correlate with the $K = |k|$ quantum number about the a and c axes in the prolate and oblate symmetric top limits respectively (see [32]).

4.6.4 Symbols for symmetry operators and labels for symmetry species

(i) Symmetry operators in space-fixed coordinates [32, 42, 47–50]

identity	E
permutation	P, p
space-fixed inversion	E^*, (P)
permutation-inversion	$P^*\,(= PE^*)$, p^*

The permutation operation P permutes the labels of identical nuclei (see [32] for notations).

Example In the NH_3 molecule, if the hydrogen nuclei are labeled 1, 2 and 3, then $P = (123)$ would symbolize the permutation where 1 is replaced by 2, 2 by 3, and 3 by 1.

The inversion operation E^* reverses the sign of all particle coordinates in the space-fixed origin, or in the molecule-fixed centre of mass if translation has been separated. It is also called the parity operator and then frequently denoted by P, although this cannot be used in parallel with permutation 'P' [32], which should then be denoted by lower case p. The convention for irreducible representations (symmetry species) still vary [32, 42, 47–50]. It is common to label symmetry with respect to space inversion by a superscript (+ for positive parity, symmetric, and − for negative parity, antisymmetric).

Non degenerate species use capital letters A, B, doubly degenerate species E, triply degenerate F, fourfold G, etc. and subscripts when needed. The symbols define species in the permutation group.

Example F_1^+ triply degenerate with respect to permutation, positive parity.

Also the mathematical notation for the symmetric group S_n of permutations is used, $\mathrm{F}_1 = [2, 1^2]$ (see [32, 49]).

(ii) Symmetry operators in molecule-fixed coordinates (Schönflies symbols) [29–31]

identity	E	rotation by $2\pi/n$	C_n
reflection	$\sigma, \sigma_\mathrm{v}, \sigma_\mathrm{d}, \sigma_\mathrm{h}$	inversion	i
rotation-reflection	$S_n\,(= C_n\sigma_\mathrm{h})$		

If C_n is the primary axis of symmetry, wavefunctions that are unchanged or change sign under the operator C_n are given species labels A or B respectively, and otherwise wavefunctions that are multiplied by $\exp(\pm 2\pi \mathrm{i} s/n)$ are given the species label E_s. Wavefunctions that are unchanged or change sign under i are labeled g (gerade) or u (ungerade) respectively. Wavefunctions that are unchanged or change sign under σ_h have species labels with a prime $'$ or a double prime $''$, respectively. For more detailed rules see [28–32].

4.6.5 Other symbols and conventions in optical spectroscopy

(i) Term symbols for atomic states

The electronic states of atoms are labeled by the value of the quantum number L for the state. The value of L is indicated by an upright capital letter: S, P, D, F, G, H, I, K, ... are used for $L =$ 0, 1, 2, 3, 4, 5, 6, 7, ..., respectively. The corresponding lower case letters are used for the orbital angular momentum of a single electron. For a many-electron atom, the electron spin multiplicity $(2S + 1)$ may be indicated as a left-hand superscript to the letter, and the value of the total angular momentum J as a right-hand subscript. If either L or S is zero only one value of J is possible, and the subscript is then usually suppressed. Finally, the electron configuration of an atom is indicated by giving the occupation of each one-electron orbital as in the examples below.

Examples B: $(1\mathrm{s})^2\,(2\mathrm{s})^2\,(2\mathrm{p})^1,\ {}^2\mathrm{P}^\circ_{1/2}$
C: $(1\mathrm{s})^2\,(2\mathrm{s})^2\,(2\mathrm{p})^2,\ {}^3\mathrm{P}_0$

A right superscript $^\circ$ may be used to indicate odd parity (negative parity −). Omission of the superscript $^\mathrm{e}$ is then to be interpreted as even parity (positive parity +).

(ii) Term symbols for molecular states

The electronic states of molecules are labeled by the symmetry species label of the wavefunction in the molecular point group. These should be Latin or Greek upright capital letters. As for atoms, the spin multiplicity $(2S + 1)$ may be indicated by a left superscript. For linear molecules the value of $\Omega\,(= \Lambda + \Sigma)$ may be added as a right subscript (analogous to J for atoms). If the value of Ω is not specified, the term symbol is taken to refer to all component states, and a right subscript r or i may be added to indicate that the components are regular (energy increases with Ω) or inverted (energy decreases with Ω) respectively.

The electronic states of molecules are also given empirical single letter labels as follows. The ground electronic state is labeled X, excited states of the same multiplicity are labeled A, B, C, ..., in ascending order of energy, and excited states of different multiplicity are labeled with lower case letters a, b, c, ... In polyatomic molecules (but not diatomic molecules) it is customary to add a tilde (e.g. $\tilde{\mathrm{X}}$) to these empirical labels to prevent possible confusion with the symmetry species label.

Finally the one-electron orbitals are labeled by the corresponding lower case letters, and the electron configuration is indicated in a manner analogous to that for atoms.

Examples The ground state of CH is $(1\sigma)^2\,(2\sigma)^2\,(3\sigma)^2\,(1\pi)^1$, X $^2\Pi_\mathrm{r}$, in which the $^2\Pi_{1/2}$ component lies below the $^2\Pi_{3/2}$ component, as indicated by the subscript r for regular.

The vibrational states of molecules are usually indicated by giving the vibrational quantum numbers for each normal mode.

(iii) Notation for spectroscopic transitions

The upper and lower levels of a spectroscopic transition are indicated by a prime $'$ and double prime $''$ respectively.

Transitions are generally indicated by giving the excited-state label, followed by the ground-state label, separated by an en–dash or an arrow to indicate the direction of the transition (emission to the right, absorption to the left).

Examples	B–A	indicates a transition between a higher energy state B and a lower energy state A;
	B→A	indicates emission from B to A;
	B←A	indicates absorption from A to B;
	$(0,2,1) \leftarrow (0,0,1)$	labels the $2\nu_2 + \nu_3 - \nu_3$ hot band in a bent triatomic molecule.

A more compact notation [51] may be used to label vibronic (or vibrational) transitions in polyatomic molecules with many normal modes, in which each vibration index r is given a superscript v_r' and a subscript v_r'' indicating the upper electronic and the lower electronic state values of the vibrational quantum number. When $v_r' = v_r'' = 0$ the corresponding index is suppressed.

Example $2_0^2\,3_1^1$ denotes the transition (0,2,1)–(0,0,1).

In order to denote transitions within the same electronic state one may use matrix notation or an arrow.

Example 2_{20} or $2_{2\leftarrow 0}$ denotes a vibrational transition within the electronic ground state from $v_2 = 0$ to $v_2 = 2$.

For rotational transitions, the value of $\Delta J = J' - J''$ is indicated by a letter labelling the branches of a rotational band: $\Delta J = -2, -1, 0, 1$, and 2 are labelled as the O-branch, P-branch, Q-branch, R-branch, and S-branch respectively. The changes in other quantum numbers (such as K for a symmetric top, or K_a and K_c for an asymmetric top) may be indicated by adding lower case letters as a left superscript according to the same rule.

Example $^\mathrm{p}$Q labels a "p-type Q-branch" in a symmetric top molecule, i.e. $\Delta K = -1$, $\Delta J = 0$.

The value of K in the lower level is indicated as a right subscript, e.g. $^\mathrm{p}\mathrm{Q}_{K''}$ or $^\mathrm{p}\mathrm{Q}_2(5)$ indicating the transition from $K'' = 2$ to $K' = 1$, the value of J'' being added in parentheses. Thus the transition in this example is specified by the quantum numbers in the lower ($''$) and upper state ($'$) as follows: $K' = 1, J' = 5 \leftarrow\ K'' = 2, J'' = 5$.

4.7 ELECTROMAGNETIC RADIATION

The quantities and symbols given here have been selected on the basis of recommendations by IUPAP [4], ISO [5.d], and IUPAC [52–55]. Terms used in photochemistry [56] have been considered as well, but definitions still vary. Terms used in high energy ionizing electromagnetic radiation, radiation chemistry, radiochemistry, and nuclear chemistry are not included or discussed.

Name	*Symbol*	*Definition*	*SI unit*	*Notes*
wavelength	λ		m	
speed of light				
in vacuum	c_0	$c_0 = 299\ 792\ 458$ m s^{-1}	m s^{-1}	1
in a medium	c	$c = c_0/n$	m s^{-1}	1
wavenumber in vacuum	$\tilde{\nu}$	$\tilde{\nu} = \nu/c_0 = 1/\lambda$	m^{-1}	2
frequency	ν	$\nu = c/\lambda$	Hz	
angular frequency	ω	$\omega = 2\pi\nu$	s^{-1}, rad s^{-1}	
refractive index	n	$n = c_0/c$	1	
radiant energy	Q, W		J	3
radiant energy density	ρ, w	$\rho = \mathrm{d}Q/\mathrm{d}V$	J m^{-3}	3
spectral radiant energy density				
in terms of frequency	ρ_ν, w_ν	$\rho_\nu = \mathrm{d}\rho/\mathrm{d}\nu$	J m^{-3} Hz^{-1}	
in terms of wavenumber	$\rho_{\tilde{\nu}}, w_{\tilde{\nu}}$	$\rho_{\tilde{\nu}} = \mathrm{d}\rho/\mathrm{d}\tilde{\nu} = \rho_\nu\, c_0$	J m^{-2}	
in terms of wavelength	ρ_λ, w_λ	$\rho_\lambda = \mathrm{d}\rho/\mathrm{d}\lambda = \rho_\nu\, c_0^2/\lambda^2$	J m^{-4}	
radiant power	P, Φ	$P = \mathrm{d}Q/\mathrm{d}t$	W	3
radiant intensity	I_e	$I_\mathrm{e} = \mathrm{d}P/\mathrm{d}\Omega$	W sr^{-1}	3, 4
radiance	L	$I_\mathrm{e} = \int L \cos\Theta\, \mathrm{d}A_\mathrm{source}$	W sr^{-1} m^{-2}	3, 4
intensity, irradiance	I, E	$I = \mathrm{d}P/\mathrm{d}A$	W m^{-2}	3, 5
spectral intensity, spectral irradiance	$I_{\tilde{\nu}}, E_{\tilde{\nu}}$	$I_{\tilde{\nu}} = \mathrm{d}I/\mathrm{d}\tilde{\nu} = c_0\, \rho_{\tilde{\nu}}$	W m^{-1}	6

(1) When there is no risk of ambiguity the subscript denoting vacuum is often omitted. n denotes the refraction index of the medium.

(2) The unit cm^{-1} is widely used for the quantity wavenumber in spectroscopy; in a medium, σ is used instead of $\tilde{\nu}$.

(3) The symbols for the quantities such as *radiant energy* and *radiant intensity* are also used for the corresponding quantities concerning visible radiation, i.e. luminous quantities and photon quantities. Subscripts e for energetic, v for visible, and p for photon may be added whenever confusion between these quantities might otherwise occur. The units used for luminous quantities are derived from the base unit candela (cd).

Examples	radiant intensity	I_e, $I_\mathrm{e} = \mathrm{d}P/\mathrm{d}\Omega$,	SI unit: W sr^{-1}
	luminous intensity	I_v,	SI unit: cd
	photon intensity	I_p,	SI unit: s^{-1} sr^{-1}

The radiant intensity I_e should be distinguished from the plain intensity or irradiance I (see note 5). Additional subscripts to distinguish absorbed (abs), transmitted (tr) or reflected (refl) quantities may be added, if necessary.

(4) The radiant intensity is the radiant power per solid angle in the direction of the point from which the source is being observed. The radiance is the radiant intensity per area of radiation source; Θ is the angle between the normal to the area element and the direction of observation as seen from the source.

(5) The intensity or irradiance is the radiation power per area that is received at a surface. Intensity, symbol I, is usually used in discussions involving collimated beams of light, as in applications of the Beer–Lambert law for spectrometric analysis. Intensity of electromagnetic radiation can also be defined as the modulus of the Poynting vector (see Section 4.3, p. 24).

(6) Spectral quantities may be defined with respect to frequency ν, wavelength λ, or wavenumber $\tilde{\nu}$; see the entry for spectral radiant energy density in this table.

Name	*Symbol*	*Definition*	*SI unit*	*Notes*
fluence	$F, (H)$	$F = \int I \mathrm{d}t = \int (\mathrm{d}P/\mathrm{d}A)\mathrm{d}t$	$\mathrm{J\ m^{-2}}$	7
resolution	$\delta\tilde{\nu}, \delta\nu$		(varies)	2, 8, 9
resolving power	R	$R = \tilde{\nu}/\delta\tilde{\nu} = \nu/\delta\nu$	1	9
free spectral range	$\Delta\tilde{\nu}$	$\Delta\tilde{\nu} = 1/2l$	$\mathrm{m^{-1}}$	2, 10
transmittance, transmission factor	τ, T	$\tau = P_{\mathrm{tr}}/P_0$	1	11, 12
absorptance, absorption factor	α	$\alpha = P_{\mathrm{abs}}/P_0$	1	11, 12
reflectance, reflection factor	ρ, R	$\rho = P_{\mathrm{refl}}/P_0$	1	11, 12
Stefan–Boltzmann constant	σ	$M_{\mathrm{bb}} = \sigma T^4$	$\mathrm{W\ m^{-2}\ K^{-4}}$	13
net absorption cross section	σ_{net}	$\sigma_{\mathrm{net}} = \kappa/N_{\mathrm{A}}$	$\mathrm{m^2}$	14
absorption coefficient				
integrated over $\tilde{\nu}$	$A, \bar{A}$	$A = \int \kappa(\tilde{\nu})\ \mathrm{d}\tilde{\nu}$	$\mathrm{m\ mol^{-1}}$	14, 15
	S	$S = A/N_{\mathrm{A}}$	m	14, 15
	$\bar{S}$	$\bar{S} = (1/pl) \int \ln(I_0/I)\ \mathrm{d}\tilde{\nu}$	$\mathrm{Pa^{-1}\ m^{-2}}$	14–16
integrated over $\ln\tilde{\nu}$	Γ	$\Gamma = \int \kappa(\tilde{\nu})\tilde{\nu}^{-1}\ \mathrm{d}\tilde{\nu}$	$\mathrm{m^2\ mol^{-1}}$	14, 15

(7) Fluence is used in photochemistry to specify the energy per area delivered in a given time interval (for instance by a laser pulse); fluence is the time integral of the fluence rate.

(8) The precise definition of resolution depends on the lineshape, but usually resolution is taken as the full line width at half maximum intensity (FWHM) on a wavenumber, $\delta\tilde{\nu}$, or frequency, $\delta\nu$, scale.

(9) This quantity characterizes the performance of a spectrometer, or the degree to which a spectral line (or a laser beam) is monochromatic.

(10) These quantities characterize a Fabry–Perot cavity, or a laser cavity. l is the cavity spacing, and $2l$ is the round-trip path length. The free spectral range is the wavenumber interval between successive longitudinal cavity modes.

(11) If scattering and luminescence can be neglected, $\tau + \alpha + \rho = 1$. In optical spectroscopy internal properties (denoted by subscript i) are defined to exclude surface effects and effects of the cuvette such as reflection losses, so that $\tau_{\mathrm{i}} + \alpha_{\mathrm{i}} = 1$, if scattering and luminescence can be neglected. This leads to the customary form of the Beer–Lambert law, $P_{\mathrm{tr}}/P_0 = I_{\mathrm{tr}}/I_0 = \tau_{\mathrm{i}} = 1 - \alpha_{\mathrm{i}} = \exp(-\kappa c l)$. Hence $A_{\mathrm{e}} = -\ln(\tau_{\mathrm{i}})$, $A_{10} = -\lg(\tau_{\mathrm{i}})$.

(12) In spectroscopy all of these quantities are commonly taken to be defined in terms of the spectral intensity, $I_{\tilde{\nu}}(\tilde{\nu})$, hence they are all regarded as functions of wavenumber $\tilde{\nu}$ (or frequency ν) across the spectrum. Thus, for example, the absorption coefficient $\alpha(\tilde{\nu})$ as a function of wavenumber $\tilde{\nu}$ defines the absorption spectrum of the sample; similarly $T(\tilde{\nu})$ defines the transmittance spectrum. Spectroscopists use $I(\tilde{\nu})$ instead of $I_{\tilde{\nu}}(\tilde{\nu})$.

(13) M_{bb} is the total emitted radiant power per area of a black body source at the same temperature.

(14) Note that these quantities give the net absorption coefficient κ, the net absorption cross section σ_{net}, and the net values of $A, S, \bar{S}, \Gamma$, and G_{net}, in the sense that they are the sums of effects due to absorption and induced emission over many spectral lines:

$$G_{\mathrm{net}}(n \leftarrow m) = \sum_{i,j} (p_i - p_j)\, G_{ji} = \left(\frac{p_m}{d_m} - \frac{p_n}{d_n}\right) \sum_{i,j} G_{ji}$$

Here p_i and p_j denote the fractional populations of states i and j ($p_i = \exp\{-E_i/kT\}/q$ in thermal equilibrium, where q is the partition function); p_m and p_n denote the corresponding fractional populations of the energy levels, and d_m and d_n the degeneracies ($p_i = p_m/d_m$, etc.).

(15) The definite integral defining these quantities may be specified by the limits of integration in parentheses, e.g. $G(\tilde{\nu}_1, \tilde{\nu}_2)$. In general the integration is understood to be taken over an absorption line or an absorption band. $A, \bar{S}$, and Γ are measures of the strength of the band in terms of amount concentration; $G_{\mathrm{net}} = \Gamma/N_{\mathrm{A}}$ and $S = A/N_{\mathrm{A}}$ are corresponding molecular quantities.

(16) The quantity $\bar{S}$ is only used for gases. Thus if $\bar{S}$ is used to report line or band intensities, the temperature should be specified. I_0 is the incident, I the transmitted intensity, thus
$\ln(I_0/I) = -\ln(I/I_0) = -\ln(1 - P_{\mathrm{abs}}/P_0) = A_{\mathrm{e}}$ (see also notes 17 and 19, p. 38).

Name	*Symbol*	*Definition*	*SI unit*	*Notes*
integrated absorption cross section	G_{ij}	$G_{ij} = \int \sigma_{ij}(\tilde{\nu})\tilde{\nu}^{-1}\,\mathrm{d}\tilde{\nu}$	m^2	14, 17
integrated net absorption cross section	G_{net}	$G_{net} = \int \sigma_{net}(\tilde{\nu})\tilde{\nu}^{-1}\,\mathrm{d}\tilde{\nu}$	m^2	14, 15
Einstein coefficient,				18, 19
spontaneous emission	A_{ij}	$\mathrm{d}N_j/\mathrm{d}t = -\sum_i A_{ij} N_j$	s^{-1}	
stimulated, induced emission, induced absorption	B_{ij}	$\mathrm{d}N_j/\mathrm{d}t = -\sum_i \rho_{\tilde{\nu}}(\tilde{\nu}_{ij}) B_{ij} N_j$	$s\ kg^{-1}$	
oscillator strength	f	$f_{ij} = \dfrac{(4\pi\varepsilon_0)\,m_e c_0}{8\pi^2 e^2}\lambda^2 A_{ij}$	1	20
absorption index	k	$k = \alpha/4\pi\tilde{\nu}$	1	21
complex refractive index	$\hat{n}$	$\hat{n} = n + \mathrm{i}k$	1	
angle of optical rotation	α		1, rad	22
optical rotatory power				
specific	$[\alpha]_\lambda{}^\theta$	$[\alpha]_\lambda{}^\theta = \alpha/\gamma l$	$rad\ m^2\ kg^{-1}$	22
molar	α_m	$\alpha_m = \alpha/cl$	$rad\ m^2\ mol^{-1}$	22, 23
(decadic) absorbance	A_{10}, A	$A_{10} = -\lg(1-\alpha_i)$	1	11,12,24
Napierian absorbance	A_e, B	$A_e = -\ln(1-\alpha_i)$	1	11,12,24
absorption coefficient,				
(linear) decadic	a, K	$a = A_{10}/l$	m^{-1}	12, 25
(linear) Napierian	α	$\alpha = A_e/l$	m^{-1}	12, 25
molar (decadic)	ε	$\varepsilon = a/c = A_{10}/cl$	$m^2\ mol^{-1}$	12,23,25,26
molar Napierian	κ	$\kappa = \alpha/c = A_e/cl$	$m^2\ mol^{-1}$	12,23,25,26

(17) G_{ij} denotes the integrated absorption cross section for the transition $j \leftarrow i$ of a single spectral line.
(18) The indices i and j refer to individual states ($E_j > E_i$, emission; $E_j < E_i$, absorption) $|E_j - E_i| = hc\tilde{\nu}_{ij}$, and $B_{ji} = B_{ij}$ in the defining equation. The coefficients B are defined here using energy density $\rho_{\tilde{\nu}}$ in terms of wavenumber, for example: $B_\nu = c_0 B_{\tilde{\nu}}$.
(19) The relation between the Einstein coefficients A and $B_{\tilde{\nu}}$ is $A = 8\pi h c_0 \tilde{\nu}^3 B_{\tilde{\nu}}$. The Einstein stimulated absorption or emission coefficient B may also be related to the transition moment $\langle i|\hat{\mu}_\rho|j\rangle$ (ρ denotes a Cartesian component) between the states i and j, or the integrated absorption cross section G_{ij}

$$B_{\tilde{\nu},ij} = \frac{8\pi^3}{3h^2 c_0\,(4\pi\varepsilon_0)} \sum_\rho |\langle i|\hat{\mu}_\rho|j\rangle|^2 = G_{ij}/h$$

(20) λ is the transition wavelength, and i and j refer to individual states; A_{ij} is the Einstein coefficient for spontaneous emission. For strongly allowed electronic transitions f is of order unity.
(21) Here α refers to the Napierian absorption coefficient. k is sometimes called 'imaginary refractive index'.
(22) The sign convention for the angle of optical rotation is as follows: α is positive if the plane of polarization is rotated clockwise as viewed looking towards the light source. If the rotation is anticlockwise, then α is negative. The optical rotation due to a solute in solution may be specified by a statement of the type

$$\alpha(589.3\ \text{nm},\ 20\ °\text{C},\ \text{sucrose},\ 10\ \text{g dm}^{-3}\ \text{in}\ H_2O,\ 10\ \text{cm path}) = +0.6647\,° = [\alpha]_{598.3\,\text{nm}}{}^{20\,°\text{C}}.$$

The same information may be conveyed by quoting either the specific optical rotatory power $\alpha/\gamma l$, or the molar optical rotatory power α/cl, where γ is the mass concentration, c is the amount (of substance) concentration, and l is the path length. For pure liquids and solids $[\alpha]_\lambda{}^\theta$ is similarly defined as $[\alpha]_\lambda{}^\theta = \alpha/\rho l$, where ρ is the mass density. Specific optical rotatory powers are customarily called *specific rotations*, and are unfortunately usually quoted without units. The absence of units may usually be taken to mean that the units are $°\ cm^3\ g^{-1}\ dm^{-1}$ for pure liquids and solutions, or $°\ cm^3\ g^{-1}\ mm^{-1}$ for solids, where ° is used as a symbol for degrees of plane angle.

4.7.1 Quantities and symbols concerned with the measurement of absorption

Net integrated absorption band intensities are usually characterized by one of the quantities $A, S, \bar{S}, \Gamma$, or G_{net} as defined in the table. The relation between these quantities is given by the (approximate) equations

$$G_{\text{net}} = \Gamma/N_{\text{A}} \approx A/(\widetilde{\nu}_0 N_{\text{A}}) \approx S/\widetilde{\nu}_0 \approx \bar{S}\,(kT/\widetilde{\nu}_0)$$

However, only the first equality is exact. The relation to A, $\bar{S}$ and S involves dividing by the band centre wavenumber $\widetilde{\nu}_0$ for a band, to correct for the fact that A, $\bar{S}$ and S are obtained by integrating over wavenumber rather than integrating over the logarithm of wavenumber used for G_{net} and Γ. This correction is only approximate for a band (although usually negligible error is involved for single-line intensities in gases). The relation to $\bar{S}$ involves the assumption that the gas is ideal (which is approximately true at low pressures), and also involves the temperature. Thus the quantities Γ and G_{net} are most simply related to more fundamental quantities such as the Einstein transition probabilities and the transition moment, and are the preferred quantities for reporting integrated line or band intensities.

Some authors even use the symbol S for any of the above quantities, particularly for any of the quantities here denoted A, S and $\bar{S}$. It is therefore particularly important to define quantities and symbols used in reporting integrated intensities.

The SI unit and commonly used units of A, S, $\bar{S}$, Γ and G, as well as numerical conversion factors are given in the table below.

Quantity	*SI unit*	*Common unit*	*Transformation coefficient*
$A, \bar{A}$	m mol^{-1}	km mol^{-1}	$(G/\text{pm}^2) \approx 16.605\,40\ \dfrac{A/(\text{km mol}^{-1})}{\widetilde{\nu}_0/\text{cm}^{-1}}$
$\bar{S}$	$\text{Pa}^{-1}\ \text{m}^{-2}$	$\text{atm}^{-1}\ \text{cm}^{-2}$	$(G/\text{pm}^2) \approx 1.362\,603 \times 10^{-2}\ \dfrac{\bar{S}/(\text{atm}^{-1}\ \text{cm}^{-2})\,(T/\text{K})}{\widetilde{\nu}_0/\text{cm}^{-1}}$
S	m	cm	$(G/\text{pm}^2) \approx 10^{20}\ \dfrac{(S/\text{cm})}{\widetilde{\nu}_0/\text{cm}^{-1}}$
Γ	$\text{m}^2\ \text{mol}^{-1}$	$\text{cm}^2\ \text{mol}^{-1}$	$(G/\text{pm}^2) \approx 1.660\,540 \times 10^{-4}\,\Gamma/(\text{cm}^2\ \text{mol}^{-1})$
G	m^2	pm^2	

Conventions for absorption intensities in condensed phases can be found in Section 2.7.2 of the Green Book.

(Notes continued)

(23) Concerning the use of "molar", see footnote 2 on p. 4.

(24) The definitions given here relate the absorbance A_{10} or A_{e} to the *internal* absorptance α_{i} (see note 11). However the subscript i on the absorptance α is often omitted. Experimental data must include corrections for reflections, scattering and luminescence, if the absorbance is to have an absolute meaning. In practice the absorbance is measured as the logarithm of the ratio of the light transmitted through a reference cell (with solvent only) to that transmitted through a sample cell.

(25) l is the absorbing path length, and c is the amount (of substance) concentration.

(26) The molar decadic absorption coefficient ε is sometimes called the "extinction coefficient". Unfortunately numerical values of the "extinction coefficient" are often quoted without specifying units; the absence of units usually means that the units are $\text{mol}^{-1}\ \text{dm}^3\ \text{cm}^{-1}$ (see also [57]). The word "extinction" should properly be reserved for the sum of the effects of absorption, scattering, and luminescence.

4.7.2 Practical equations linking A_{fi}, G_{fi} and M_{fi} as well as f_{fi} [32]

The indices f and i are running indices and seem to be useful in the context given below.

$$\begin{aligned} A_{fi} &= \frac{8\pi}{c_0^2}\,\nu_{fi}^3\,G_{fi} \\ G_{fi} &= \frac{8\pi^3}{3hc_0(4\pi\epsilon_0)}\,|M_{fi}|^2 \\ (G_{fi}/\mathrm{pm}^2) &\approx 41.624\,|M_{fi}/\mathrm{D}|^2 \end{aligned}$$

where 1 D (debye) $\approx 3.335\,641 \times 10^{-30}$ C m and

$$|M_{fi}|^2 = \sum_\rho |\langle f|\widehat{\mu}_\rho|i\rangle|^2$$

$\langle f|\widehat{\mu}_\rho|i\rangle$ is the Cartesian component of the transition dipole moment between states f and i (see Section 4.6.1, p. 30) in the electric dipole approximation, and $\widetilde{\nu}_{fi}$ is the transition wavenumber between these states.

In electronic spectroscopy one also often uses the 'oscillator strength', a quantity of dimension one (number):

$$f_{fi} = \frac{(4\pi\epsilon_0)\,m_\mathrm{e}c_0}{8\pi^2e^2}\,\lambda_{fi}^2\,A_{fi} \approx 1.4992\times 10^{-14}\left(\frac{A_{fi}}{\mathrm{s}^{-1}}\right)\left(\frac{\lambda_{fi}}{\mathrm{nm}}\right)^2$$

4.8 SOLID STATE

See also [58] and the *International Tables for Crystallography,* Volume A [59].

Name	*Symbol*	*Definition*	*SI unit*	*Notes*
Bravais lattice vector	$\boldsymbol{t}$		m	
fundamental translation vector for the crystal lattice	$\boldsymbol{a}_1; \boldsymbol{a}_2; \boldsymbol{a}_3,$ $\boldsymbol{a}; \boldsymbol{b}; \boldsymbol{c}$	$\boldsymbol{t} = n_1\boldsymbol{a}_1 + n_2\boldsymbol{a}_2 + n_3\boldsymbol{a}_3$	m	1
(angular) fundamental translation vectors for the reciprocal lattice	$\boldsymbol{b}_1; \boldsymbol{b}_2; \boldsymbol{b}_3,$ $\boldsymbol{a}^*; \boldsymbol{b}^*; \boldsymbol{c}^*$	$\boldsymbol{a}_i \cdot \boldsymbol{b}_k = 2\pi\delta_{ik}$	m^{-1}	2
(angular) reciprocal lattice vector	$\boldsymbol{G}$	$\boldsymbol{G} = h_1\boldsymbol{b}_1 + h_2\boldsymbol{b}_2 + h_3\boldsymbol{b}_3$	m^{-1}	
unit cell lengths	$a; b; c$		m	1
unit cell angles	$\alpha; \beta; \gamma$		rad, 1	1
reciprocal unit cell lengths	$a^*; b^*; c^*$		m^{-1}	
reciprocal unit cell angles	$\alpha^*; \beta^*; \gamma^*$		rad, 1	
fractional coordinates	$x; y; z$	$x = X/a$	1	3
atomic scattering factor	f		1	
structure factor (indices h, k, l)	$F(h, k, l)$	$F = \sum_{n=1}^{N} f_n \mathrm{e}^{2\pi\,\mathrm{i}(hx_n + ky_n + lz_n)}$	1	4
lattice plane spacing	d, d_{hkl}		m	
Bragg angle	θ	$\lambda = 2d_{hkl}\sin\theta$	rad, 1	5
position vector	$\boldsymbol{r}_j, \boldsymbol{R}_j$		m	6
equilibrium position vector (ion)	$\boldsymbol{R}_0$		m	
displacement vector (ion)	$\boldsymbol{u}$	$\boldsymbol{u} = \boldsymbol{R} - \boldsymbol{R}_0$	m	
Debye–Waller factor	B, D	$D = \mathrm{e}^{-2\langle(\boldsymbol{q}\cdot\boldsymbol{u})^2\rangle}$	1	7
Debye angular wavenumber	$q_\mathrm{D}, k_\mathrm{D}$	$k_\mathrm{D} = (C_\mathrm{i}\, 6\pi^2)^{1/3}$	m^{-1}	8
Debye angular frequency	ω_D	$\omega_\mathrm{D} = k_\mathrm{D}\, c_0$	s^{-1}	8
Debye frequency	ν_D	$\nu_\mathrm{D} = \omega_\mathrm{D}/2\pi$	s^{-1}	
Debye wavenumber	$\tilde{\nu}_\mathrm{D}$	$\tilde{\nu}_\mathrm{D} = \nu_\mathrm{D}/c_0$	m^{-1}	8
Debye temperature	Θ_D	$\Theta_\mathrm{D} = h\nu_\mathrm{D}/k_\mathrm{B}$	K	
Fermi energy	$E_\mathrm{F}, \varepsilon_\mathrm{F}$	$\varepsilon_\mathrm{F} = \lim_{T\to 0} \mu$	J	9
(electron) work function	Φ	$\Phi = E_\infty - E_\mathrm{F}$	J	10

(1) n_1, n_2 and n_3 are integers. The unit cell lengths a, b and c are the norms of $\boldsymbol{a}$, $\boldsymbol{b}$ and $\boldsymbol{c}$, respectively. Together with the unit cell angles α, β and γ they are called the lattice constants or lattice parameters.

(2) These are sometimes defined by $\boldsymbol{a}_i \cdot \boldsymbol{b}_k = \delta_{ik}$ (Kronecker delta δ_{ik}, see Section 1.8, p. 8).

(3) X denotes the coordinate of dimension length.

(4) N is the number of atoms in the unit cell and h, k, l are integers.

(5) θ is the glancing angle of the incident beam of wavelength λ and hkl denotes the set of diffracting planes.

(6) Electron and ion position vectors are denoted by $\boldsymbol{r}$ and $\boldsymbol{R}$. The subscript j relates to particle j.

(7) $\hbar\boldsymbol{q}$ is the momentum transfer in neutron diffraction, $\langle\,\rangle$ denotes thermal averaging. $\boldsymbol{q}$ is the angular wave vector.

(8) C_i is the ion density, c_0 is the speed of light in vacuum.

(9) The commonly used unit for this quantity is eV. μ is the chemical potential per entity.

(10) E_∞ is the electron energy at rest at infinite distance [60].

4.8.1 Symbols for planes and directions in crystals

Miller indices of a crystal face, or of a single net plane	(hkl) or $(h_1h_2h_3)$
indices of the Bragg reflection from the set of parallel net planes (hkl)	hkl or $h_1h_2h_3$
indices of a set of all symmetrically equivalent crystal faces, or net planes	$\{hkl\}$ or $\{h_1h_2h_3\}$
lattice point in the periodic space	uvw
indices of a lattice direction (zone axis)	$[uvw]$
indices of a set of symmetrically equivalent lattice directions	$\langle uvw\rangle$

The variables u, v and w are integer numbers; they denote coordinates of a primitive lattice point in the periodic space. As such, they can define any direction from the origin of the lattice. The direction vector starting at the origin ends at that lattice point uvw.

For a single plane or crystal face, or a specific direction, a negative number is indicated by a bar over the number.

Example $(\bar{1}10)$ denotes the single net plane or a crystal face $h = -1$, $k = 1$, $l = 0$.

4.9 STATISTICAL THERMODYNAMICS

Name	*Symbol*	*Definition*	*SI unit*	*Notes*
Boltzmann constant	k, k_B		$\mathrm{J\ K^{-1}}$	
(molar) gas constant	R	$R = N_\mathrm{A}\, k$	$\mathrm{J\ K^{-1}\ mol^{-1}}$	1
molecular velocity vector	$\boldsymbol{c}, \boldsymbol{u}, \boldsymbol{v}$	$\boldsymbol{c} = \mathrm{d}\boldsymbol{r}/\mathrm{d}t$	$\mathrm{m\ s^{-1}}$	2
molecular momentum vector	$\boldsymbol{p}$	$\boldsymbol{p} = m\boldsymbol{c}$	$\mathrm{kg\ m\ s^{-1}}$	2
velocity distribution function	$f(c_x)$	$f = \left(\dfrac{m}{2\pi kT}\right)^{1/2} \exp\left(-\dfrac{m{c_x}^2}{2kT}\right)$	$\mathrm{m^{-1}\ s}$	2
speed distribution function	$F(c)$	$F = 4\pi c^2 \left(\dfrac{m}{2\pi kT}\right)^{3/2} \exp\left(-\dfrac{mc^2}{2kT}\right)$	$\mathrm{m^{-1}\ s}$	2
average speed	$\bar{c}, \bar{u}, \bar{v}, \langle c\rangle, \langle u\rangle, \langle v\rangle$	$\bar{c} = \int cF(c)\, \mathrm{d}c$	$\mathrm{m\ s^{-1}}$	
generalized coordinate	q		(varies)	3
generalized momentum	p	$p = \partial L/\partial \dot{q}$	(varies)	3
volume in phase space	Ω	$\Omega = (1/h) \int p\, \mathrm{d}q$	1	
probability	P, p		1	
statistical weight, degeneracy	g, d, W, ω, β		1	4
(cumulative) number of states	W, N	$W(E) = \sum_i \mathrm{H}(E - E_i)$	1	5, 6
density of states	$\rho(E)$	$\rho(E) = \mathrm{d}W(E)/\mathrm{d}E$	$\mathrm{J^{-1}}$	
partition function, sum over states,				
canonical ensemble	Q, Z, q	$Q = \sum_i g_i \exp(-E_i/kT)$	1	6
microcanonical ensemble	Ω, z, Z	$Z = \sum_{i=1}^{Z} 1$	1	
symmetry number	σ, s		1	
reciprocal energy parameter	β	$\beta = 1/kT$	$\mathrm{J^{-1}}$	
density operator	$\widehat{\rho}$, $\widehat{\sigma}$	$\widehat{\rho} = \sum_k p_k \vert\Psi_k\rangle\langle\Psi_k\vert$	1	7
density matrix element	P_{mn}, ρ_{mn}	$P_{mn} = \langle\phi_m\vert\widehat{\rho}\vert\phi_n\rangle$	1	8
statistical entropy	S	$S = -k_\mathrm{B} \sum_i p_i \ln p_i$	$\mathrm{J\ K^{-1}}$	9

(1) N_A is the Avogadro constant (Section 4.10, p. 44). Concerning the name "molar gas constant", see footnote 2 on p. 4.
(2) $\boldsymbol{r}$ is the molecular position vector and m is the mass of the molecule. Cartesian vector components can be given as r_x, r_y, r_z, c_x, c_y, c_z or p_x, p_y, p_z.
(3) If q is a length then p is a momentum. In the definition of p, L denotes the Lagrangian.
(4) β is sometimes used for a spin statistical weight and the degeneracy is also called polytropy.
(5) $\mathrm{H}(x)$ is the Heaviside function, W or $W(E)$ is the number of quantum states with energy less than E.
(6) E_i, g_i denote the energy and degeneracy respectively of the ith level of the quantum system.
(7) $\vert\Psi_k\rangle$ refers to the quantum state k of the system and p_k to the probability of this state in an ensemble. If $p_k = 1$ for a given state k one speaks of a pure state, otherwise of a mixture.
(8) The density matrix $\boldsymbol{P}$ is defined by its matrix elements P_{mn} in a set of basis states $\vert\phi_m\rangle$.
(9) The equilibrium (maximum) entropy for a microcanonical ensemble results then as $S = k_\mathrm{B} \ln Z$.

4.10 GENERAL CHEMISTRY

Name	*Symbol*	*Definition*	*SI unit*	*Notes*
number of entities	N		1	1
Avogadro constant	N_A, L		mol^{-1}	2
amount of substance	n	$n = N/N_A$	mol	2
amount of substance B, amount of B	$n_B, n(B)$	$n_B = N_B/L$	mol	2, 3
atomic mass	m_a		kg	
mass of entity B	$m_B, m(B)$		kg	3
atomic mass constant	m_u	$m_u = m_a(^{12}C)/12$	kg	4
molar mass of entity B	M_B	$M_B = m_B/n_B$	$kg\ mol^{-1}$	3, 5, 6
molar mass constant	M_u	$M_u = m_u N_A$	$kg\ mol^{-1}$	5, 7
relative molecular mass, (relative molar mass, molecular weight)	$M_{r,B}$	$M_{r,B} = m_B/m_u$	1	5, 7
relative atomic mass, (atomic weight)	A_r	$A_r = m_a/m_u$	1	7
molar volume of entity B	$V_{m,B}$	$V_{m,B} = V_B/n_B$	$m^3\ mol^{-1}$	3, 5, 6
mass fraction	w	$w_B = m_B/\sum_j m_j$	1	3
volume fraction	ϕ	$\phi_B = V_B/\sum_j V_j$	1	3, 8
amount-of-substance fraction, amount fraction, (mole fraction)	x_B, y_B	$x_B = n_B/\sum_j n_j$	1	3, 5, 9
(total) pressure	$p, (P)$		Pa	
partial pressure	p_B	$p_B = y_B\, p$	Pa	3
mass concentration	γ, ρ	$\gamma_B = m_B/V$	$kg\ m^{-3}$	3, 10
number concentration	C	$C = N/V$	m^{-3}	3, 10
(amount) concentration	c	$c = n/V$	$mol\ m^{-3}$	3, 10, 11
(amount) concentration of B	c_B, [B]	$c_B = n_B/V$	$mol\ m^{-3}$	3, 10, 11

(1) Entities may be atoms, molecules, ions, electrons, other particles, or specified groups of such particles.
(2) The 'Avogadro number' [3] is the numerical value of the Avogadro constant when expressed in mol^{-1}, thus $\{N_A\} = 6.022\,140\,76 \times 10^{23}$. The symbol L is used to avoid confusion with number of entity A.
(3) The definition applies to a specific entity B which should always be indicated by a subscript or in parentheses, e.g. n_B or $n(B)$; N_B refers to the number of a specific entity B. When more entities are involved one uses B_j and a summation index j runs over all entities B_j, e.g. $m_j = m(B_j)$. When the chemical composition is written out, parentheses should be used, $n(O_2)$.

Example When the amount of O_2 is equal to 3 mol, $n(O_2) = 3$ mol, the amount of $(1/2)\ O_2$ is equal to 6 mol, and $n((1/2)\ O_2) = 6$ mol. Thus $n((1/2)\ O_2) = 2\, n(O_2)$.

(4) m_u is equal to the unified atomic mass unit, with symbol u, i.e. $m_u = 1$ u. The dalton, symbol Da, is used as an alternative name for the unified atomic mass unit.
(5) Concerning the names 'mole fraction', 'molar mass', 'molar volume' etc. see footnote 2, p. 4.
(6) The definition applies to pure substance B, where m_B is its mass and V_B is its volume.
(7) For historical reasons the terms "molecular weight" and "atomic weight" are still used. For molecules M_r is the relative molecular mass or "molecular weight". For atoms M_r is the relative atomic mass or "atomic weight", and the symbol A_r may be used. M_r may also be called the relative molar mass, $M_{r,B} = M_B/M_u$, where $M_u \approx 1\ g\ mol^{-1}$.
(8) Here, V_B and V_i are the volumes of appropriate components prior to mixing [20].
(9) For condensed phases x is used, and for gaseous mixtures y may be used [61,62].
(10) V is the volume of the mixture. It should be noted that in polymer science, the symbol c is used for "mass concentration" instead of γ and the wording "concentration" is often used by omitting "mass". Neither is recommended because confusion can arise outside polymer science.

Name	*Symbol*	*Definition*	*SI unit*	*Notes*
solubility	s	$s_B = c_B$ (saturated solution)	mol m^{-3}	2
molality	m, b	$m_B = n_B/m_A$	mol kg^{-1}	2, 3, 5, 12
stoichiometric number	ν_j		1	13
extent of reaction	ξ	$\xi = (n_j - n_{j,0})/\nu_j$	mol	14
degree of reaction	α	$\alpha = \xi/\xi_{max}$	1	14

(Notes continued)
(11) Amount concentration is sometimes called *molarity.* Units commonly used for amount concentration are mol L^{-1} (or mol dm^{-3}), mmol L^{-1}, µmol L^{-1} etc., often denoted M, mM, µM etc. (pronounced molar, see also Section 1.4 and footnote 2, p. 4, millimolar, micromolar). Thus M is often treated as a symbol for mol L^{-1} . IFCC and IUPAC Chemistry and Human Health Division prefer "substance concentration" instead of "amount concentration".
(12) In this definition the symbol m is used with two different meanings: m_B denotes the *molality* of solute B, m_A denotes the *mass* of solvent A (thus the unit mol kg^{-1}). This confusion of notation is avoided by using the symbol b for molality, see also footnote 2, p. 4.
(13) A general chemical equation can be written as

$$0 = \sum_j \nu_j \, B_j$$

where B_j denotes a species in the reaction and ν_j the corresponding stoichiometric number (negative for reactants and positive for products). The ammonia synthesis is equally well expressed in these two possible ways:

(i) $(1/2)\ N_2 + (3/2)\ H_2 = NH_3$ $\nu(N_2) = -1/2,$ $\nu(H_2) = -3/2,$ $\nu(NH_3) = +1$
(ii) $N_2 + 3\ H_2 = 2\ NH_3$ $\nu(N_2) = -1,$ $\nu(H_2) = -3,$ $\nu(NH_3) = +2$

The stoichiometric numbers ν_j are defined through the stoichiometric equation. They are negative for reactants and positive for products. The values of the stoichiometric numbers depend on how the reaction equation is written. $n_{B,0}$ denotes the value of n_B at "zero time", when $\xi = 0$ mol.

Examples The formation of hydrogen bromide from the elements can equally well be written in either of these two ways
$(1/2)\ H_2 + (1/2)\ Br_2 = HBr$ or $H_2 + Br_2 = 2\ HBr$

(14) The changes $\Delta n_j = n_j - n_{j,0}$ in the amounts of any reactant and product j during the course of the reaction is governed by one parameter, the extent of reaction ξ, through the equation

$$n_j = n_{j,0} + \nu_j \, \xi$$

The extent of reaction depends on how the reaction is written, but it is independent of which entity in the reaction is used in the definition. Thus, for reaction (i), note 12, when $\xi = 2$ mol, then $\Delta n(N_2) = -1$ mol, $\Delta n(H_2) = -3$ mol and $\Delta n(NH_3) = +2$ mol. For reaction (ii), when $\Delta n(N_2) = -1$ mol then $\xi = 1$ mol. ξ_{max} is the value of ξ when at least one of the reactants is exhausted.

4.10.1 Other symbols and conventions in chemistry

(i) The symbols for the chemical elements

The symbols for the chemical elements are (in most cases) derived from their Latin names and consist of one or two letters which should always be printed in roman (upright) type.

Examples I, U, Pa, C

The symbols have two different meanings (which also reflects on their use in chemical formulae and equations):

(a) On a *microscopic* level they can denote an atom of the element. For example, Cl denotes a chlorine atom having 17 protons and 18 or 20 neutrons (giving a mass number of 35 or 37), the difference being ignored. Its mass is on average 35.45 Da in terrestrial samples.

(b) On a *macroscopic* level they denote a sample of the element. For example, Fe denotes a sample of iron, and He a sample of helium. They may also be used as a shorthand to denote the element: "Fe is one of the most common elements in the Earth's crust."

The term *nuclide* implies an atom of specified atomic number (proton number) and mass number (nucleon number). A nuclide may be specified by attaching the mass number as a left superscript to the *symbol* for the element, as in ^{14}C, or added after the *name* of the element, as in carbon–14. Nuclides having the same atomic number but different mass numbers are called isotopic nuclides or *isotopes*, as in ^{12}C, ^{14}C. If no left superscript is attached, the symbol is read as including all isotopes in natural abundance: $n(\mathrm{Cl}) = n(^{35}\mathrm{Cl}) + n(^{37}\mathrm{Cl})$. Nuclides having the same mass number but different atomic numbers are called isobaric nuclides or *isobars*: ^{14}C, ^{14}N. The atomic number may be attached as a left subscript: $^{14}_{6}C$, $^{14}_{7}N$.

The ionic charge number is denoted by a right superscript.

Examples	Na^+	a sodium(1+) ion (cation)
	$^{79}Br^-$	a [^{79}Br]bromide(1−) ion (anion)
	Al^{3+}	an aluminium(3+) ion (cation)

Excited electronic states may be denoted by an asterisk.

Examples H^*, Cl^*

Oxidation states (oxidation numbers) are denoted by positive or negative roman numerals or by zero in chemical names (in parenthesis) or in chemical formulae (as a right superscript).

Examples Mn^{VII}, manganese(VII), O^{-II}, Ni^0

The positions and meanings of indices around the symbol of the element are summarized as follows:

left superscript	mass number
left subscript	atomic number
right superscript	charge number, oxidation state/oxidation number, excitation
right subscript	number of atoms per entity

(ii) Chemical formulae

As in the case of chemical symbols of elements, chemical formulae have two different meanings:

(a) On a *microscopic* level they denote one atom, one molecule, one ion, one radical, etc. The number of atoms in an entity (always an integer) is indicated by a right subscript, the numeral 1 being omitted. Groups of atoms may be enclosed in parentheses. Charge numbers of ions and excitation symbols are added as right superscripts to the formula. The radical nature of an entity may be expressed by adding a dot to the symbol. The nomenclature for radicals, ions, radical ions, and related species is described in [63, 64].

Examples	Xe, N_2, C_6H_6, Na^+, $SO_4{}^{2-}$	
	$(CH_3)_3COH$	(a 2-methylpropan-2-ol molecule)
	$NO_2{}^*$	(an excited nitrogen dioxide molecule)
	NO	(a nitrogen oxide molecule)
	$NO^\bullet$	(a nitrogen oxide molecule, stressing its free radical character)

In writing the formula for a more complex ion, spacing for charge number may be added (staggered arrangement), which is the preferred writing [65], as well as parentheses and brackets: $SO_4{}^{2-}$, $(SO_4)^{2-}$, $[SO_4]^{2-}$.

Specific electronic states of entities (atoms, molecules, ions) can be denoted by giving the electronic term symbol in parentheses. Vibrational and rotational states can be specified by giving the corresponding quantum numbers.

Examples	$Hg(^3P_1)$	a mercury atom in the triplet-P-one state
	$HF(v = 2, J = 6)$	a hydrogen fluoride molecule in the vibrational state $v = 2$ and the rotational state $J = 6$
	$H_2O^+(^2A_1)$	a dihydridooxygen(1+) ion in the doublet-A-one state

(b) On a *macroscopic* level a formula denotes a sample of a chemical substance (not necessarily stable, or capable of existing in isolated form). The chemical composition is denoted by right subscripts (not necessarily integers; the numeral 1 being omitted). A "formula unit" (which is *not* a unit of a quantity!) is an entity specified as a group of atoms (see (iii) and (iv) below).

Examples Na, Na^+, NaCl, $Fe_{0.91}S$, $XePtF_6$

The formula is used in expressions like $\rho(H_2SO_4)$ to indicate the mass density of sulfuric acid, although the formula H_2SO_4 does not specify a particular state of sulfuric acid. When specifying amount of substance the formula is often multiplied with a factor, normally a small integer or a fraction, see examples in (iv) and (v). Chemical formulae may be written in different ways according to the information they convey.

(iii) Equations for chemical reactions

(a) On a *microscopic* level the reaction equation represents an elementary reaction (an event involving single atoms, molecules, and radicals), or the sum of a set of such reactions. Stoichiometric numbers are ± 1 (sometimes ± 2). A single arrow is used to connect reactants and products in an elementary reaction. An equal sign is used for the "net" reaction, the result of a set of elementary reactions (see Section 4.12.1, p. 57).

$H + Br_2 \rightarrow HBr + Br$	one elementary step in HBr formation
$H_2 + Br_2 = 2\ HBr$	the sum of several such elementary steps

(b) On a *macroscopic* level, different symbols are used connecting the reactants and products in the reaction equation, with the following meanings:

$H_2 + Br_2 = 2\ HBr$	stoichiometric equation
$H_2 + Br_2 \rightarrow 2\ HBr$	net forward reaction
$H_2 + Br_2 \leftrightarrows 2\ HBr$	reaction, both directions
$H_2 + Br_2 \rightleftharpoons 2\ HBr$	equilibrium

The two-sided arrow $\leftrightarrow$ should not be used for reactions to avoid confusion with resonance structures.

Redox equations are often balanced with integer stoichiometric numbers (first equation in *Examples*) or per one exchanged electron (second equation in *Examples*):

Examples

$$2\ KMn^{VII}O_4 + 16\ HCl = 2\ Mn^{II}Cl_2 + 5\ Cl_2 + 2\ KCl + 8\ H_2O$$
$$(1/5)\ KMn^{VII}O_4 + (8/5)\ HCl = (1/5)\ Mn^{II}Cl_2 + (1/2)\ Cl_2 + (1/5)\ KCl + (4/5)\ H_2O$$

(iv) Amount of substance and the specification of entities

The quantity "amount of substance" or "chemical amount" ("Stoffmenge" in German, "quantité de matière" in French) has been used by chemists for a long time without a proper name. It was simply referred to as the "number of moles". This practice should be abandoned; the name of a physical quantity should not contain the name of a unit (few would use "number of metres" as a synonym for "length"), see also footnote 2 on p. 4.

The amount of substance is proportional to the number of specified elementary entities of that substance; the proportionality constant is the same for all substances and is the reciprocal of the Avogadro constant. The elementary entities may be chosen as convenient, not necessarily as physically real individual particles. In the examples below, (1/2) $MgSO_4$, (1/5) $KMnO_4$, etc. are artificial in the sense that no such elementary entities exist. Since the amount of substance and all physical quantities derived from it depend on this choice, it is essential to specify the entities to avoid ambiguities.

Examples	$n(\mathrm{Cl})$, n_{Cl}	amount of Cl, amount of chlorine atoms
	$n(\mathrm{Cl_2})$	amount of Cl_2, amount of chlorine molecules
	$n(\mathrm{H_2SO_4})$	amount of H_2SO_4
	$n((1/5)\ \mathrm{KMnO_4})$	amount of (1/5) $KMnO_4$
	$M(\mathrm{P_4})$	molar mass of tetraphosphorus P_4
	$c(\mathrm{Cl^-})$, $[\mathrm{Cl^-}]$, $c_{\mathrm{Cl^-}}$	amount concentration of Cl^-
	$\Lambda(\mathrm{MgSO_4})$	molar conductivity of $MgSO_4$
	$\Lambda((1/2)\ \mathrm{MgSO_4})$	molar conductivity of (1/2) $MgSO_4$
	$\lambda(\mathrm{Mg^{2+}})$	ionic conductivity of Mg^{2+}
	$\lambda((1/2)\ \mathrm{Mg^{2+}})$	ionic conductivity of (1/2) Mg^{2+}

Using definitions of various quantities we can derive equations like

$$n((1/5)\ \mathrm{KMnO_4}) = 5n(\mathrm{KMnO_4})$$
$$\lambda((1/2)\ \mathrm{Mg^{2+}}) = (1/2)\lambda(\mathrm{Mg^{2+}})$$
$$[(1/2)\ \mathrm{H_2SO_4}] = 2\ [\mathrm{H_2SO_4}]$$

Note that "amount of sulfur" is an ambiguous statement because it might imply $n(\mathrm{S})$, $n(\mathrm{S_8})$, or $n(\mathrm{S_2})$, etc. In most cases analogous statements are less ambiguous. Thus for compounds the implied entity is usually the molecule or the common formula entity, and for solid metals it is the atom.

Examples	"2 mol of water" implies $n(\mathrm{H_2O}) = 2$ mol
	"0.5 mol of sodium chloride" implies $n(\mathrm{NaCl}) = 0.5$ mol
	"3 mmol of iron" implies $n(\mathrm{Fe}) = 3$ mmol

Such statements should be avoided whenever there might be ambiguity.

In the equation $pV = nRT$, the entity implied in the definition of n should be an independently translating particle (a whole molecule for a gas), whose nature is unimportant.

(v) States of aggregation

The following one-, two- or three-letter symbols are used to represent the states of aggregation of chemical species. The letters are appended in parenthesis to the formula or symbol, and should be printed in roman (upright) type without a full stop (period).

a, ads	species adsorbed on a surface	g	gas or vapor
aq	aqueous solution	l	liquid
aq, ∞	aqueous solution at infinite dilution	s	solid
cr	crystalline	sln	solution
f	fluid phase		

Examples	HCl(g)	hydrogen chloride in the gaseous state
	NaOH(aq)	aqueous solution of sodium hydroxide
	$\Delta_{\mathrm{f}}H^{\ominus}(\mathrm{H_2O, l})$	standard enthalpy of formation of liquid water

The symbols g, l, etc. to denote gas phase, liquid phase, etc., are also sometimes used as a right superscript, and the Greek letter symbols α, β, etc. may be similarly used to denote phase α, phase β, etc. in a general notation.

Examples	$V_{\mathrm{m}}{}^{\mathrm{l}}$, $V_{\mathrm{m}}{}^{\mathrm{s}}$ molar volume of the liquid phase l and of the solid phase s
	$S_{\mathrm{m}}{}^{\alpha}$, $S_{\mathrm{m}}{}^{\beta}$ molar entropy of phase α and of phase β

4.11 CHEMICAL THERMODYNAMICS

The names and symbols of the more generally used quantities given here are also recommended by IUPAP [4] and by ISO [5.c,5.e]. Additional information can be found in [1.d,1.j] and [62].

Name	*Symbol*	*Definition*	*SI unit*	*Notes*
heat	Q, q		J	1
work	W, w		J	1
internal energy	U	$\mathrm{d}U = đQ + đW$	J	1
enthalpy	H	$H = U + pV$	J	
Celsius temperature	θ, t	$\theta/°\mathrm{C} = T/\mathrm{K} - 273.15$	°C	2, 3
entropy	S	$\mathrm{d}S = đQ_{\mathrm{rev}}/T$	$\mathrm{J\ K^{-1}}$	
Helmholtz energy	A, F	$A = U - TS$	J	
Gibbs energy	G	$G = H - TS$	J	
molar quantity X	X_{m}, $(\overline{X})$	$X_{\mathrm{m}} = X/n$	$[X]\ \mathrm{mol^{-1}}$	3–5
specific quantity X	x	$x = X/m$	$[X]\ \mathrm{kg^{-1}}$	4, 5
heat capacity,				
at constant pressure	C_p	$C_p = (\partial H/\partial T)_p$	$\mathrm{J\ K^{-1}}$	
at constant volume	C_V	$C_V = (\partial U/\partial T)_V$	$\mathrm{J\ K^{-1}}$	
ratio of heat capacities	γ, (κ)	$\gamma = C_p/C_V$	1	
Joule–Thomson coefficient	μ, μ_{JT}	$\mu = (\partial T/\partial p)_H$	$\mathrm{K\ Pa^{-1}}$	
thermal power	Φ, P	$\Phi = \mathrm{d}Q/\mathrm{d}t$	W	
virial coefficient,				
second	B	$pV_{\mathrm{m}} = RT(1 + B/V_{\mathrm{m}} +$	$\mathrm{m^3\ mol^{-1}}$	6
third	C	$C/V_{\mathrm{m}}^2 + \cdots)$	$\mathrm{m^6\ mol^{-2}}$	6
van der Waals	a	$(p + a/V_{\mathrm{m}}^2)(V_{\mathrm{m}} - b) = RT$	$\mathrm{J\ m^3\ mol^{-2}}$	
coefficients	b		$\mathrm{m^3\ mol^{-1}}$	
compression factor	Z	$Z = pV_{\mathrm{m}}/RT$	1	
partial molar quantity	X_{B}, $(\overline{X}_{\mathrm{B}})$	$X_{\mathrm{B}} = (\partial X/\partial n_{\mathrm{B}})_{T,p,n_{j\neq\mathrm{B}}}$	$[X]\ \mathrm{mol^{-1}}$	3, 7
chemical potential,	μ	$\mu_{\mathrm{B}} = (\partial G/\partial n_{\mathrm{B}})_{T,p,n_{j\neq\mathrm{B}}}$	$\mathrm{J\ mol^{-1}}$	8
standard	$\mu^{\ominus}$, μ°		$\mathrm{J\ mol^{-1}}$	8, 9
absolute activity	λ	$\lambda_{\mathrm{B}} = \exp(\mu_{\mathrm{B}}/RT)$	1	8
(relative) activity	a	$a_{\mathrm{B}} = \exp\left(\dfrac{\mu_{\mathrm{B}} - \mu_{\mathrm{B}}^{\ominus}}{RT}\right)$	1	8–10

(1) In the differential form, đ denotes an inexact differential. The given equation in integrated form is $\Delta U = Q + W$. $Q > 0$ and $W > 0$ indicate an increase in the energy of the system.
(2) T is the absolute or thermodynamic temperature. The quantity Celsius temperature is sometimes misnamed "centigrade temperature", see also footnote 2 on p. 4.
(3) Concerning the terms "molar quantity", "Celsius temperature" etc. see footnote 2, p. 4.
(4) The definition applies to pure substance. However, the concept of molar and specific quantities may also be applied to mixtures. n is the amount of substance, and m the mass.
(5) X is an extensive quantity, whose SI unit is $[X]$. The entities should be specified for molar quantities.
(6) Another set of "pressure virial coefficients" may be defined by expanding pV_{m} in powers of p.
(7) The symbol applies to entities B which should be specified. The bar may be used to distinguish partial molar X from X when necessary.

Example The partial molar volume of Na_2SO_4 in aqueous solution may be denoted $\overline{V}(Na_2SO_4, \mathrm{aq})$ to distinguish it from the volume of the solution $V(Na_2SO_4, \mathrm{aq})$.

(8) The definition applies to entities B which should be specified. The chemical potential can be defined equivalently by the corresponding partial derivatives of other thermodynamic functions (U, H, A).
(9) The symbol $^{\ominus}$ or $^{\circ}$ is used to indicate standard which must be specified. They are equally acceptable (for definitions of standard states, see Section 4.11.1 (iv), p. 52).

Name	*Symbol*	*Definition*	*SI unit*	*Notes*
standard partial molar enthalpy	$H_B^\ominus$	$H_B^\ominus = \mu_B^\ominus + TS_B^\ominus$	J mol^{-1}	3, 8, 9
standard reaction Gibbs energy	$\Delta_r G^\ominus$	$\Delta_r G^\ominus = \sum_B \nu_B \mu_B^\ominus$	J mol^{-1}	9, 11, 12
affinity of reaction	A, $\mathcal{A}$	$A = -(\partial G/\partial \xi)_{p,T} = -\sum_B \nu_B \mu_B$	J mol^{-1}	12
standard reaction enthalpy	$\Delta_r H^\ominus$	$\Delta_r H^\ominus = \sum_B \nu_B H_B^\ominus$	J mol^{-1}	9, 11, 12
standard reaction entropy	$\Delta_r S^\ominus$	$\Delta_r S^\ominus = \sum_B \nu_B S_B^\ominus$	J mol^{-1} K^{-1}	9, 11, 12
reaction quotient	Q	$Q = \prod_B a_B{}^{\nu_B}$	1	13
equilibrium constant,	$K^\ominus, K$	$K^\ominus = \exp(-\Delta_r G^\ominus/RT)$	1	9, 11, 12, 14
pressure basis	K_p	$K_p = \prod_B p_B{}^{\nu_B}$	Pa$^{\Sigma\nu_B}$	12, 15
concentration basis	K_c	$K_c = \prod_B c_B{}^{\nu_B}$	(mol m^{-3})$^{\Sigma\nu_B}$	12, 15
molality basis	K_m	$K_m = \prod_B m_B{}^{\nu_B}$	(mol kg^{-1})$^{\Sigma\nu_B}$	3, 12, 15
fugacity	$f, \tilde{p}$	$f_B = \lambda_B \lim_{p\to 0}(p_B/\lambda_B)_T$	Pa	8
fugacity coefficient	ϕ	$\phi_B = f_B/p_B$	1	
Henry's law constant	k_H	$k_{H,B} = \lim_{x_B\to 0}(f_B/x_B)$	Pa	8, 16
activity coefficient				
referenced to Raoult's law	f	$f_B = a_B/x_B$	1	8, 17
referenced to Henry's law				
molality basis	γ_m	$a_{m,B} = \gamma_{m,B} m_B/m^\ominus$	1	3, 8, 18
concentration basis	γ_c	$a_{c,B} = \gamma_{c,B} c_B/c^\ominus$	1	8, 18
amount fraction basis	γ_x	$a_{x,B} = \gamma_{x,B} x_B$	1	7, 17
ionic strength,				
molality basis	I_m, I	$I_m = (1/2)\sum_i m_i z_i{}^2$	mol kg^{-1}	3
concentration basis	I_c, I	$I_c = (1/2)\sum_i c_i z_i{}^2$	mol m^{-3}	

(10) In this equation, the pressure dependence of the activity has been neglected as is often done for condensed phases at atmospheric pressure. The definition of $\mu^\ominus$ depends on the choice of the standard state.
(11) The symbol r indicates reaction in general. In particular cases r can be replaced by another appropriate subscript, e.g. $\Delta_f H^\ominus$ denotes the standard molar enthalpy of formation. Δ_r can be interpreted as operator symbol $\Delta_r \stackrel{\text{def}}{=} \partial/\partial\xi$.
(12) The reaction must be specified for which this quantity applies.
(13) This quantity applies in general to a system which is not in equilibrium.
(14) This quantity is equal to Q at chemical equilibrium, when $\Delta_r G$ is zero. It is dimensionless and its value depends on the choice of standard state, which must be specified. The symbol $K^\ominus$ and the name "standard equilibrium constant" is recommended [62]. Many chemists prefer the symbol K and the name "thermodynamic equilibrium constant".
(15) These quantities are not in general dimensionless. One can define in an analogous way an equilibrium constant in terms of fugacity K_f, etc. The equilibrium constant of dissolution of an electrolyte (describing the equilibrium between excess solid phase and solvated ions) is often called a solubility product, denoted K_{sol} or K_s (or $K_{sol}{}^\ominus$ or $K_s{}^\ominus$ as appropriate). Similarly, the equilibrium constant for an acid dissociation is K_a, for base hydrolysis K_b, and for water dissociation K_w. p_B, c_B, and m_B are equilibrium values.
(16) Henry's law is sometimes expressed in terms of molalities or concentration and then the corresponding units of Henry's law constant are Pa kg mol^{-1} or Pa m^3 mol^{-1}, respectively.
(17) This quantity applies to pure phases, substances in mixtures, or solvents.
(18) This quantity applies to solutes.

4.11.1 Other symbols and conventions in chemical thermodynamics

A more extensive description of this subject can be found in [62].

(i) Symbols used as subscripts to denote a physical chemical process or reaction
These symbols should be printed in roman (upright) type, without a full stop (period).

adsorption	ads	mixing of fluids	mix
atomization	at	reaction in general	r
combustion reaction	c	solution (of solute in solvent)	sol
dilution (of a solution)	dil	sublimation (solid $\rightarrow$ gas)	sub
displacement	dpl	transition (between two phases)	trs
immersion	imm	triple point	tp
melting, fusion (solid $\rightarrow$ liquid)	fus	vaporization, evaporation (liquid $\rightarrow$ gas)	vap

(ii) Recommended superscripts

activated complex, transition state	‡, ≠	infinite dilution	∞
apparent	app	pure substance	*
excess quantity	E	standard	$\ominus$, o
ideal	id		

(iii) Examples of the use of the symbol Δ
The symbol Δ denotes a change in an extensive thermodynamic quantity for a process. The addition of a subscript to the Δ denotes a change in the property.

Examples $\Delta_{\text{vap}}H = H_{\text{m}}(\text{g}) - H_{\text{m}}(\text{l})$ for the molar enthalpy of vaporization.
$\Delta_{\text{vap}}H = 40.7\ \text{kJ mol}^{-1}$ for water at 100 °C under its own vapor pressure.
This can also be written ΔH_{vap}, but this usage is not recommended.

The subscript r is used to denote changes associated with a *chemical reaction*. Symbols such as $\Delta_{\text{r}}H$ are defined by the equation

$$\Delta_{\text{r}}H = \sum_{\text{B}} \nu_{\text{B}} H_{\text{B}} = (\partial H/\partial \xi)_{T,p}$$

It is thus essential to specify the stoichiometric reaction equation when giving numerical values for such quantities in order to define the extent of reaction ξ and the value of the stoichiometric numbers ν_{B} .

Example $N_2(\text{g}) + 3\ H_2(\text{g}) = 2\ NH_3(\text{g})$, $\Delta_{\text{r}}H^{\ominus}(298.15\ \text{K}) = -92.4\ \text{kJ mol}^{-1}$
$\Delta_{\text{r}}S^{\ominus}(298.15\ \text{K}) = -199\ \text{J mol}^{-1}\ \text{K}^{-1}$

The mol^{-1} in the units identifies the quantities in this example as the change per extent of reaction. They may be called the molar enthalpy and entropy of reaction, and a subscript m may be added to the symbol to emphasize the difference from the integral quantities if desired.

The *standard reaction quantities* are particularly important. They are defined by the equations

$$\Delta_{\text{r}}H^{\ominus} = \sum_{\text{B}} \nu_{\text{B}} H_{\text{B}}{}^{\ominus} \quad \text{and} \quad \Delta_{\text{r}}S^{\ominus} = \sum_{\text{B}} \nu_{\text{B}} S_{\text{B}}{}^{\ominus} \quad \text{and} \quad \Delta_{\text{r}}G^{\ominus} = \sum_{\text{B}} \nu_{\text{B}} \mu_{\text{B}}{}^{\ominus}$$

It is important to specify notation with care for these symbols. The relation to the affinity of the reaction is

$$-A = \Delta_{\text{r}}G = \Delta_{\text{r}}G^{\ominus} + RT \ln \left(\prod_{\text{B}} a_{\text{B}}{}^{\nu_{\text{B}}} \right) = \Delta_{\text{r}}G^{\ominus} + RT \ln Q$$

and the relation to the standard equilibrium constant is $\Delta_r G^\ominus = -RT \ln K^\ominus$. The product of the activities is the reaction quotient Q, see also p. 50.

The term *combustion* and symbol c denote the complete oxidation of a substance. For the definition of complete oxidation of substances containing elements other than C, H and O see [66]. The corresponding reaction equation is written so that the stoichiometric number ν of the substance is -1.

Example The standard enthalpy of combustion of gaseous methane is $\Delta_c H^\ominus(CH_4, g, 298.15\ K) = -890.3\ kJ\ mol^{-1}$, implying the reaction $CH_4(g) + 2\ O_2(g) = CO_2(g) + 2\ H_2O(l)$.

The term *formation* and symbol f denote the formation of the substance from elements in their reference state (usually the most stable state of each element at the chosen temperature and standard pressure). The corresponding reaction equation is written so that the stoichiometric number ν of the substance is $+1$. Similarly, the term *atomization*, symbol at, denotes a process in which a substance is separated into its constituent atoms in the ground state in the gas phase. The corresponding reaction equation is written so that the stoichiometric number ν of the substance is -1.

(iv) Standard states [1.j] and [62]

The standard chemical potential of substance B at temperature $T, \mu_B{}^\ominus(T)$, is the value of the chemical potential under standard conditions, specified as follows. Three differently defined standard states are recognized.

For a gas phase. The standard state for a gaseous substance, whether pure or in a gaseous mixture, is the (hypothetical) state of the pure substance B in the gaseous phase at the standard pressure $p = p^\ominus$ and exhibiting ideal gas behavior. The standard chemical potential is defined as

$$\mu_B{}^\ominus(T) = \lim_{p \to 0} [\mu_B(T, p, y_B, ...) - RT \ln(y_B p / p^\ominus)]$$

For a pure phase, or a mixture, or a solvent, in the liquid or solid state. The standard state for a liquid or solid substance, whether pure or in a mixture, or for a solvent, is the state of the pure substance B in the liquid or solid phase at the standard pressure $p = p^\ominus$. The standard chemical potential is defined as

$$\mu_B(T) = \mu_B{}^*(T, p^\ominus)$$

For a solute in solution. For a solute in a liquid or solid solution the standard state is referenced to the ideal dilute behavior of the solute. It is the (hypothetical) state of solute B at the standard molality $m^\ominus$, standard pressure $p^\ominus$, and behaving like the infinitely dilute solution. The standard chemical potential is defined as

$$\mu_B{}^\ominus(T) = [\mu_B(T, p^\ominus, m_B, ...) - RT \ln(m_B/m^\ominus)]^\infty$$

The chemical potential of the solute B as a function of the molality m_B at constant pressure $p = p^\ominus$ is then given by the expression

$$\mu_B(m_B) = \mu_B{}^\ominus + RT \ln(m_B \gamma_{m,B}/m^\ominus)$$

Sometimes (amount) concentration c is used as a variable in place of molality m; both of the above equations then have c in place of m throughout. Occasionally amount fraction x is used in place of m; both of the above equations then have x in place of m throughout, and $x^\ominus = 1$. Although the standard state of a solute is always referenced to ideal dilute behavior, the definition of the standard state and the value of the standard chemical potential $\mu^{\ominus}$ are different depending on whether molality m, concentration c, or amount fraction x is used as a variable.

(v) Standard pressures, molality, and concentration

In principle one may choose any values for the standard pressure $p^{\ominus}$, the standard molality $m^{\ominus}$, and the standard concentration $c^{\ominus}$, although the choice must be specified. For example, in tabulating data appropriate to high pressure chemistry it may be convenient to choose a value of $p^{\ominus} = 100$ MPa ($= 1$ kbar).

In practice, however, the most common choice is

$$\begin{aligned} p^{\ominus} &= 0.1\ \text{MPa} = 100\ \text{kPa}\ (= 1\ \text{bar}) \\ m^{\ominus} &= 1\ \text{mol kg}^{-1} \\ c^{\ominus} &= 1\ \text{mol dm}^{-3} \end{aligned}$$

These values for $m^{\ominus}$ and $c^{\ominus}$ are universally accepted. The value for $p^{\ominus} = 100$ kPa, is the IUPAC recommendation since 1982 [1.j] for tabulating thermodynamic data and should be specified. The conversion of values corresponding to different $p^{\ominus}$ is described in [67–69]. The newer value of $p^{\ominus}$, 100 kPa is sometimes called the *standard state pressure*.

(vi) Biochemical standard states

Special standard states that are close to physiological conditions are often chosen. The biochemical standard state is often chosen at $[H^+] = 10^{-7}$ mol dm^{-3}. The concentrations of the solutes may be grouped together as for example, the total phosphate concentration rather than the concentration of each component, (H_3PO_4, $H_2PO_4{}^-$, $HPO_4{}^{2-}$, $PO_4{}^{3-}$), separately. Standard and other reference states must be specified with care [70, 71].

(vii) Thermodynamic properties

Values of many thermodynamic quantities represent basic chemical properties of substances and serve for further calculations. Extensive tabulations exist, e.g. [72–76].

(viii) Reference state (of an element)

The state in which the element is stable at the chosen standard state pressure and for a given temperature [21].

4.12 CHEMICAL KINETICS AND PHOTOCHEMISTRY

The recommendations given here are based on previous IUPAC recommendations [1.c,1.k] and [22], which are not in complete agreement. Recommendations regarding photochemistry are given in [56] and for recommendations on reporting of chemical kinetics data see also [77]. A glossary of terms used in chemical kinetics has been given in [78].

Name	*Symbol*	*Definition*	*SI unit*	*Notes*
rate of change of quantity X	$\dot{X}$	$\dot{X} = \mathrm{d}X/\mathrm{d}t$	(varies with X)	1
rate of conversion	$\dot{\xi}$	$\dot{\xi} = \mathrm{d}\xi/\mathrm{d}t$	$\mathrm{mol\ s^{-1}}$	2
rate of concentration change (due to chemical reaction)	$r_\mathrm{B}, v_\mathrm{B}$	$r_\mathrm{B} = \mathrm{d}c_\mathrm{B}/\mathrm{d}t$	$\mathrm{mol\ m^{-3}\ s^{-1}}$	3
rate of reaction (based on amount concentration)	v, v_c	$v = \nu_\mathrm{B}{}^{-1}\mathrm{d}c_\mathrm{B}/\mathrm{d}t$ $= \dot{\xi}/V$	$\mathrm{mol\ m^{-3}\ s^{-1}}$	2, 3
rate of reaction (based on number concentration), (reaction rate)	v, v_C	$v_C = \nu_\mathrm{B}{}^{-1}\mathrm{d}C_\mathrm{B}/\mathrm{d}t$	$\mathrm{m^{-3}\ s^{-1}}$	2, 3
partial order of reaction	m_B	$v = k\prod_\mathrm{B} c_\mathrm{B}{}^{m_\mathrm{B}}$	1	4
overall order of reaction	m	$m = \sum_\mathrm{B} m_\mathrm{B}$	1	
rate constant, rate coefficient	$k, k(T)$	$v = k\prod_\mathrm{B} c_\mathrm{B}{}^{m_\mathrm{B}}$	$(\mathrm{m^3\ mol^{-1}})^{m-1}\ \mathrm{s^{-1}}$	5
rate constant of unimolecular reaction	$k_\mathrm{uni}, k_\mathrm{uni}(T, c_\mathrm{M})$	$v = k_\mathrm{uni} c_\mathrm{B}$	$\mathrm{s^{-1}}$	6
at high pressure	k_∞	$k_\mathrm{uni}(c_\mathrm{M} \to \infty)$	$\mathrm{s^{-1}}$	6
at low pressure	k_0	$v = k_0 c_\mathrm{M} c_\mathrm{B}$	$\mathrm{m^3\ mol^{-1}\ s^{-1}}$	6

(1) E.g. rate of pressure change $\dot{p} = \mathrm{d}p/\mathrm{d}t$, for which the SI unit is Pa $\mathrm{s^{-1}}$, rate of entropy change $\mathrm{d}S/\mathrm{d}t$ with SI unit J $\mathrm{K^{-1}\ s^{-1}}$.
(2) The reaction must be specified, for which this quantity applies, by giving the stoichiometric equation.
(3) Note that the rate of concentration change r_B with entity B and v can also be defined on the basis of partial pressure, number concentration, surface concentration, etc., with analogous definitions. If necessary differently defined rates of reaction can be distinguished by a subscript, e.g. $v_p = \nu_\mathrm{B}{}^{-1}\mathrm{d}p_\mathrm{B}/\mathrm{d}t$, etc. Note that the rate of reaction (also called reaction rate) can only be defined for a reaction of known and time-independent stoichiometry, in terms of a specified reaction equation; also the second equation for the rate of reaction follows from the first only if the volume V is constant and c_B is uniform throughout (more generally, the definition applies to local concentrations). The derivatives must be those due to the chemical reaction considered; in open systems, such as flow systems, effects due to input and output processes must be taken into account separately, as well as transport processes.
(4) The symbol applies to entity B. The order of reaction is only defined if the particular rate law applies. m is a real number.
(5) The temperature dependent rate constant k or $k(T)$ and pre-exponential factors A are usually quoted in either $(\mathrm{dm^3\ mol^{-1}})^{m-1}\ \mathrm{s^{-1}}$ or on a molecular scale in $(\mathrm{cm^3})^{m-1}\ \mathrm{s^{-1}}$ or $(\mathrm{cm^3\ molecule^{-1}})^{m-1}\ \mathrm{s^{-1}}$. Note that "molecule" is not a unit, but is often included for clarity, although this does not conform to accepted usage. Rate constants are frequently quoted as decadic logarithms.

Example second order reaction $k = 10^{8.2}\ \mathrm{cm^3\ mol^{-1}\ s^{-1}}$ or $\lg(k/\mathrm{cm^3\ mol^{-1}\ s^{-1}}) = 8.2$

(6) The rates of unimolecular reactions show a dependence upon the concentration c_M of a collision partner M.

Name	*Symbol*	*Definition*	*SI unit*	*Notes*
half life	$t_{1/2}$	$c(t_{1/2}) = c(0)/2$	s	
relaxation time, lifetime, mean life	τ	$\Delta c(\tau) = \Delta c(0)/\mathrm{e}$	s	7
(Arrhenius) activation energy	$E_\mathrm{A}, E_\mathrm{a}$	$E_\mathrm{A} = RT^2\mathrm{d}(\ln k)/\mathrm{d}T$	J mol^{-1}	8
pre-exponential factor, frequency factor	$A, A(T)$	$A = k\exp(E_\mathrm{A}/RT)$	$(\mathrm{m^3\ mol^{-1}})^{m-1}\ \mathrm{s^{-1}}$	8
hard sphere radius	r		m	
collision diameter	d_AB	$d_\mathrm{AB} = r_\mathrm{A} + r_\mathrm{B}$	m	
collision cross section	σ	$\sigma = \pi {d_\mathrm{AB}}^2$	m^2	9
mean relative speed between A and B	$\bar{c}_\mathrm{AB}$	$\bar{c}_\mathrm{AB} = (8k_\mathrm{B}T/\pi\mu)^{1/2}$	m s^{-1}	10
collision frequency				
of A with A	$z_\mathrm{A}(\mathrm{A})$	$z_\mathrm{A}(\mathrm{A}) = \sqrt{2}\, C_\mathrm{A}\sigma\bar{c}$	s^{-1}	9
of A with B	$z_\mathrm{A}(\mathrm{B})$	$z_\mathrm{A}(\mathrm{B}) = C_\mathrm{B}\sigma\bar{c}_\mathrm{AB}$	s^{-1}	9
collision density, collision number				
of A with A	Z_AA	$Z_\mathrm{AA} = C_\mathrm{A} z_\mathrm{A}(\mathrm{A})$	s^{-1} m^{-3}	11
of A with B	Z_AB	$Z_\mathrm{AB} = C_\mathrm{A} z_\mathrm{A}(\mathrm{B})$	s^{-1} m^{-3}	11
collision frequency factor	z_AB	$z_\mathrm{AB} = Z_\mathrm{AB}/N_\mathrm{A} c_\mathrm{A} c_\mathrm{B}$	m^3 mol^{-1} s^{-1}	11
scattering matrix	$\boldsymbol{S}$		1	12
transition probability	P_{ji}	$P_{ji} = \lvert S_{ji}\rvert^2$	1	12
standard enthalpy of activation	$\Delta^\ddagger H^\circ$	$k(T) = \dfrac{kT}{h}\exp\left(\dfrac{\Delta^\ddagger S^\circ}{R}\right)\exp\left(\dfrac{-\Delta^\ddagger H^\circ}{RT}\right)$	J mol^{-1}	13
volume of activation	$\Delta^\ddagger V^\circ$	$\Delta^\ddagger V^\circ = -RT(\partial(\ln k)/\partial p)_T$	m^3 mol^{-1}	

(7) τ is defined as the time interval in which a concentration perturbation Δc falls to 1/e of its initial value $\Delta c(0)$. This lifetime or decay time must be distinguished from the half life.

(8) One may use as defining equation

$$E_\mathrm{A} = -R\mathrm{d}(\ln k)/\mathrm{d}(1/T)$$

The term Arrhenius activation energy is to be used only for the empirical quantity defined in the table. Other empirical equations with different "activation energies" are also being used. E_A and A can depend on temperature.

(9) The collision cross section σ is a constant in the hard sphere collision model, but generally it is energy dependent. One may furthermore define a temperature dependent average collision cross section (see note 16). C_i denotes the number concentration of entity i.

(10) μ is the reduced mass, $\mu = m_\mathrm{A} m_\mathrm{B}/(m_\mathrm{A} + m_\mathrm{B})$.

(11) Z_AA and Z_AB are the total number of AA or AB collisions per time and volume in a system containing only A molecules, or containing molecules of two types, A and B. Three-body collisions can be treated in a similar way. N_A is the Avogadro constant.

(12) The first index refers to the final and the second to the initial channel. i and j denote reactant and product channels, respectively. The scattering matrix $\boldsymbol{S}$ is used in quantum scattering theory [79]. $\boldsymbol{S}$ is a unitary matrix $\boldsymbol{S}\boldsymbol{S}^\dagger = \mathbf{1}$. $P_{ji} = \lvert S_{ji}\rvert^2$ is the probability that collision partners incident in channel i will emerge in channel j.

(13) The quantities $\Delta^\ddagger H^\circ$, the standard internal energy of activation $\Delta^\ddagger U^\circ$, the standard entropy of activation $\Delta^\ddagger S^\circ$, and $\Delta^\ddagger G^\circ$, etc. are used in the transition state theory of chemical reactions. They are appropriately used only in connection with elementary reactions. It is not recommended to omit the standard symbol $^\circ$, which would introduce ambiguity. The relation between the rate constant k and these quantities of activation is more generally

Name	*Symbol*	*Definition*	*SI unit*	*Notes*
standard Gibbs energy of activation	$\Delta^{\ddagger}G^{\circ}$, $\Delta G^{\ddagger}$	$\Delta^{\ddagger}G^{\circ} = \Delta^{\ddagger}H^{\circ} - T\Delta^{\ddagger}S^{\circ}$	$\mathrm{J\ mol^{-1}}$	13
molecular partition function for transition state	$q^{\ddagger}, q^{\ddagger}(T)$	$k_{\infty} = \kappa \dfrac{k_{\mathrm{B}}T}{h} \dfrac{q^{\ddagger}}{q_{\mathrm{A}}} \exp(-E_0/k_{\mathrm{B}}T)$	1	13
transmission coefficient	κ, γ	$\kappa = \dfrac{k_{\infty}h}{k_{\mathrm{B}}T} \dfrac{q_{\mathrm{A}}}{q^{\ddagger}} \exp(E_0/k_{\mathrm{B}}T)$	1	13
molecular partition function per volume	$\tilde{q}$	$\tilde{q} = q/V$	$\mathrm{m^{-3}}$	13
specific rate constant of bimolecular reaction at collision energy E_{t}	$k_{\mathrm{bi}}(E_{\mathrm{t}})$	$k_{\mathrm{bi}}(E_{\mathrm{t}}) = \sigma(E_{\mathrm{t}})\sqrt{2E_{\mathrm{t}}/\mu}$	$\mathrm{m^3\ s^{-1}}$	10
density of states	$\rho(E, J, \ldots)$	$\rho(E, J, \ldots) = \mathrm{d}N(E, J, \ldots)/\mathrm{d}E$	$\mathrm{J^{-1}}$	
number (sum) of states	$N(E)$, $W(E)$	$N(E) = \int_0^E \rho(E')\,\mathrm{d}E'$	1	
Michaelis constant	K_{M}, K_{m}	$v = \dfrac{V\,[\mathrm{S}]}{K_{\mathrm{M}} + [\mathrm{S}]}$	$\mathrm{mol\ m^{-3}}$	14
rate (coefficient) matrix	$\boldsymbol{K}$, K_{fi}	$-\dfrac{\mathrm{d}\boldsymbol{c}}{\mathrm{d}t} = \boldsymbol{K}\,\boldsymbol{c}$	$\mathrm{s^{-1}}$	15
quantum yield, photochemical yield	Φ, ϕ		1	16
fluorescence rate constant	k_{f}	$\dfrac{\mathrm{d}[h\nu]_{\mathrm{f}}}{\mathrm{d}t} = k_{\mathrm{f}}\,c^{*}$	$\mathrm{s^{-1}}$	17, 18
natural lifetime	τ_0	$\tau_0 = 1/k_{\mathrm{f}}$	s	17
natural linewidth	Γ, Γ_{f}	$\Gamma = \hbar k_{\mathrm{f}}$	J	17
predissociation linewidth	Γ_{p}, Γ_{diss}	$\Gamma_{\mathrm{p}} = \hbar k_{\mathrm{p}}$	J	17

(13) (continued)

$$k = \kappa(k_{\mathrm{B}}T/h)\exp(-\Delta^{\ddagger}G^{\circ}/RT)$$

where k has the dimensions of a first-order rate constant. A rate constant for a reaction of order n is obtained by multiplication with $(c^{\circ})^{1-n}$. For a bimolecular reaction of an ideal gas one multiplies by $V^{\circ} = (c^{\circ})^{-1} = (kT/p^{\circ})$. κ is a transmission coefficient, and $\Delta^{\ddagger}G^{\circ} = \Delta^{\ddagger}H^{\circ} - T\,\Delta^{\ddagger}S^{\circ}$. The choice of p° and c° in general affects the values of $\Delta^{\ddagger}H^{\circ}$, $\Delta^{\ddagger}S^{\circ}$ and $\Delta^{\ddagger}G^{\circ}$.

The statistical mechanical formulation of transition state theory results in the equation for k_{∞} as given in the table for a unimolecular reaction in the high pressure limit, and for a bimolecular reaction one has (often with $\kappa = 1$ assumed)

$$k_{\mathrm{bi}} = \kappa \frac{k_{\mathrm{B}}T}{h} \frac{\tilde{q}^{\ddagger}}{\tilde{q}_{\mathrm{A}}\tilde{q}_{\mathrm{B}}} \exp(-E_0/k_{\mathrm{B}}T)$$

where $q^{\ddagger}$ is the partition function of the transition state and $\tilde{q}^{\ddagger}, \tilde{q}_{\mathrm{A}}, \tilde{q}_{\mathrm{B}}$ are partition functions per volume for the transition state and the reaction partners A and B. E_0 is the threshold energy for reaction, below which no reaction is assumed to occur. In transition state theory, it is the difference of the zero-point level of the transition state and the zero-point level of reactants.

(14) The Michaelis constant arises in the treatment of a mechanism of enzyme catalysis

$$\mathrm{E} + \mathrm{S} \underset{k_{-1}}{\overset{k_1}{\rightleftarrows}} (\mathrm{ES}) \underset{k_{-2}}{\overset{k_2}{\rightleftarrows}} \mathrm{E} + \mathrm{P}$$

where E is the enzyme, S the substrate, ES the enzyme-substrate complex and P the product. V is called limiting rate [78].

(15) In generalized first-order kinetics, the rate equation can be written as a matrix equation, with the concentration vector $\boldsymbol{c} = (c_1, c_2, \ldots, c_n)^{\mathsf{T}}$ and the first-order rate coefficients K_{fi} as matrix elements.

4.12.1 Other symbols, terms, and conventions used in chemical kinetics

Additional descriptions can be found in [78].

(i) Elementary reactions

Reactions that occur at the molecular level in one step are called *elementary reactions.* It is conventional to define them as unidirectional, written with a simple arrow always from left to right. The number of relevant reactant particles on the left hand side is called the *molecularity* of the elementary reaction.

Examples	$\mathrm{A \rightarrow B + C}$	unimolecular reaction (also monomolecular)
	$\mathrm{A + B \rightarrow C + D}$	bimolecular reaction
	$\mathrm{A + B + C \rightarrow D + E}$	trimolecular reaction (also termolecular)

(ii) Composite mechanisms

A reaction that involves more than one elementary reaction is said to occur by a composite mechanism. The terms complex mechanism, indirect mechanism, and stepwise mechanism are also commonly used. Special types of mechanisms include chain-reaction mechanisms, catalytic reaction mechanisms, etc.

Examples A simple mechanism is composed of forward and reverse reactions

$$\mathrm{A \rightarrow B + C}$$

$$\mathrm{B + C \rightarrow A}$$

It is in this particular case conventional to write these in one line

$$\mathrm{A \rightleftarrows B + C}$$

However, it is useful in kinetics to distinguish this from a net reaction (at equilibrium), which is written either with two one-sided arrows or an "equal" sign

$$\mathrm{A \rightleftharpoons B + C}$$

$$\mathrm{A = B + C}$$

(Notes continued)

(16) The quantum yield ϕ is defined in general by the *number of defined events* divided by *number of photons absorbed* [56]. For a photochemical reaction it can be defined as

$$\phi = \left| \frac{\mathrm{d}\xi/\mathrm{d}t}{\mathrm{d}n_\gamma/\mathrm{d}t} \right|$$

which is the *rate of conversion* divided by the *rate of photon absorption.*

(17) For exponential decay by spontaneous emission (fluorescence) one has for the decay of the excited state concentration c^*

$$\frac{\mathrm{d}[h\nu]}{\mathrm{d}t} = -\frac{\mathrm{d}c^*}{\mathrm{d}t} = k_\mathrm{f} c^*$$

The full width at half maximum Γ of a Lorentzian absorption line is related to the rate constant of fluorescence. For a predissociation Γ_p is related to the predissociation rate constant k_p. However, linewidths may also have other contributions in practice.

(18) The einstein is sometimes either used as the unit of amount of substance of photons, where one einstein corresponds to 1 mol of photons, or as the unit of energy where one einstein corresponds to the energy $N_\mathrm{A} h\nu$ of 1 mol of monochromatic photons with frequency ν.

When one combines a composite mechanism to obtain a net reaction, one should not use the simple arrow in the resulting equation.

Example

$\mathrm{A \rightarrow B + C}$	unimolecular elementary reaction
$\mathrm{B + C \rightarrow D + E}$	bimolecular elementary reaction
$\mathrm{A = D + E}$	net reaction (no elementary reaction, no molecularity)

It is furthermore useful to distinguish the stoichiometric equation defining the reaction rate and rate constant from the equation defining the elementary reaction and rate law.

Example Recombination of methyl radicals in the high pressure limit
The elementary reaction is $\mathrm{CH_3 + CH_3 \rightarrow C_2H_6}$
This has a second order rate law. If one uses the stoichiometric equation
$\mathrm{2\,CH_3 = C_2H_6}$
the definition of the reaction rate gives

$$v_C = -\frac{1}{2}\frac{\mathrm{d}C_{\mathrm{CH_3}}}{\mathrm{d}t} = k\,{C_{\mathrm{CH_3}}}^2$$

Similar to reaction enthalpies, rate coefficients are only defined with a given stoichiometric equation. It is recommended always to state explicitly the stoichiometric equation and the differential equation for the reaction rate to avoid ambiguities. If no other statement is made, the stoichiometric coefficients should be as in the reaction equation (for the elementary reaction). However, it is not always possible to use a reaction equation as stoichiometric equation.

Example The trimolecular reaction for hydrogen atoms
The elementary reaction is $\mathrm{H + H + H \rightarrow H_2 + H}$
This has a third order rate law. The stoichiometric equation is
$\mathrm{2\,H = H_2}$
and the definition of the reaction rate gives

$$v_c = -\frac{1}{2}\frac{\mathrm{d[H]}}{\mathrm{d}t} = k\,\mathrm{[H]}^3$$

The bimolecular reaction $\mathrm{H + H \rightarrow H_2}$ with the same stoichiometric equation is a *different* elementary reaction (a very unlikely one).
Other examples include catalyzed reactions such as $\mathrm{X + A \rightarrow X + B}$ with the stoichiometric equation $\mathrm{A = B}$ and the rate $v_c = -\mathrm{d[A]}/\mathrm{d}t = k\mathrm{[A][X]}$ ($\neq \mathrm{d[X]}/\mathrm{d}t = 0$). Again, the unimolecular reaction $\mathrm{A \rightarrow B}$ exists, but would be a *different* elementary reaction with the same stoichiometry.
In the rate laws for catalyzed reactions the amount or the concentration of a catalyst may appear, which does not appear in the stoichiometric equation of the reaction. When the amount of a catalyst cannot be expressed as a number of elementary entities, an amount of substance, or a mass, a 'catalytic activity' can still be defined as a property of a catalyst measured by a catalyzed rate of conversion under specified optimized conditions. One uses commonly the quantity 'catalytic activity' with the SI derived unit mol $\mathrm{s^{-1}}$, the name 'katal' and the symbol kat.

4.13 ELECTROCHEMISTRY

Electrochemical concepts, terminology and symbols are more extensively described in [1.i] and [80–84].

Name	*Symbol*	*Definition*	*SI unit*	*Notes*
Faraday constant	F	$F = eN_{\mathrm{A}}$	$\mathrm{C\ mol^{-1}}$	
charge number of an ion	z	$z_{\mathrm{B}} = Q_{\mathrm{B}}/e$	1	1
ionic strength,				
molality basis	I_m, I	$I_m = (1/2) \sum_i m_i {z_i}^2$	$\mathrm{mol\ kg^{-1}}$	
concentration basis	I_c, I	$I_c = (1/2) \sum_i c_i {z_i}^2$	$\mathrm{mol\ m^{-3}}$	2
mean ionic molality	$m_\pm$	$m_\pm{}^{(\nu_+ + \nu_-)} = m_+{}^{\nu_+} m_-{}^{\nu_-}$	$\mathrm{mol\ kg^{-1}}$	3
mean ionic activity coefficient	$\gamma_\pm$	$\gamma_\pm{}^{(\nu_+ + \nu_-)} = \gamma_+{}^{\nu_+} \gamma_-{}^{\nu_-}$	1	3
mean ionic activity	$a_\pm$	$a_\pm = m_\pm \gamma_\pm / m^\circ$	1	3–5
activity of an electrolyte	$a(\mathrm{A}_{\nu_+}\mathrm{B}_{\nu_-})$	$a(\mathrm{A}_{\nu_+}\mathrm{B}_{\nu_-}) = a_\pm{}^{(\nu_+ + \nu_-)}$	1	3
pH	pH	$\mathrm{pH} = -\lg(a_{\mathrm{H^+}})$	1	5
Galvani potential difference	$\Delta\phi$	$\Delta_\alpha{}^\beta \phi = \phi^\beta - \phi^\alpha$	V	6
electrochemical potential	$\tilde{\mu}_{\mathrm{B}}{}^\alpha$	$\mu_{\mathrm{B}}{}^{\alpha\circ} + RT \ln a_{\mathrm{B}} + z_{\mathrm{B}} F \phi^\alpha$	$\mathrm{J\ mol^{-1}}$	1, 7
electrode potential	E, U		V	8
cell potential	$E_{\mathrm{cell}}, U_{\mathrm{cell}}$	$E_{\mathrm{cell}} = E_{\mathrm{R}} - E_{\mathrm{L}}$	V	9
standard electrode potential	E°	$E^\circ = -\Delta_{\mathrm{r}} G^\circ / zF$	V	10, 11
electron number of an electrochemical reaction (charge number)	z, n	$z = \lvert \nu_{\mathrm{e}} \rvert$	1	12

(1) The definition applies to entities B, e is the elementary charge.
(2) To avoid confusion with the cathodic current, symbol I_{c} (note roman subscript), the symbol I or sometimes μ (when the current is denoted by I) is used for ionic strength based on concentration.
(3) ν_+ and ν_- are the numbers of cations and anions per formula unit of an electrolyte $\mathrm{A}_{\nu+}\mathrm{B}_{\nu-}$.

Example For $\mathrm{Al_2(SO_4)_3}$, $\nu_+ = 2$ and $\nu_- = 3$.

m_+ and m_-, and γ_+ and γ_-, are the molalities and activity coefficients of cations and anions. If the molality of $\mathrm{A}_{\nu+}\ \mathrm{B}_{\nu-}$ is m, then $m_+ = \nu_+ m$ and $m_- = \nu_- m$. A similar definition is used on a concentration scale for the mean ionic concentration $c_\pm$.
(4) $m^\circ = 1\ \mathrm{mol\ kg^{-1}}$.
(5) For an individual ion, neither the activity a_+, a_- nor the activity coefficient γ_+, γ_- is experimentally measurable. The definition of pH is discussed in the Green Book (Section 2.13.1(viii), p. 75 in [20]).
(6) $\Delta\phi$ is the inner potential difference between points within the bulk phases α and β.
(7) The electrochemical potential of ion B in a phase α is the partial molar Gibbs energy of the ion.
(8) The absolute value of the electrode potential cannot be measured, so E is always reported relative to the potential of some reference electrode, e.g. that of a standard hydrogen electrode (SHE) (see Section 4.13.1 (vi), p. 62).
(9) This quantity is the potential difference of an electrochemical cell. The two electrodes of the electrochemical cell can be distinguished by the subscripts L (left) and R (right) (see Section 4.13.1 (iii), p. 61). E_{L} and E_{R} are the electrode potentials of these two electrodes. In electrochemical engineering, the cell potential is exclusively denoted by U.
(10) The symbols $^\circ$ and $^\circ$ are both used to indicate standard state; they are equally acceptable.
(11) The standard potential is the value of the *equilibrium potential* of an electrode under standard conditions. $\Delta_{\mathrm{r}} G^\circ$ is the standard Gibbs energy of this electrode reaction, written as a reduction, with respect to that of the standard hydrogen electrode.
(12) z (or n) is the number of electrons in the balanced electrode reaction as written. It is a positive integer, equal to $\lvert \nu_{\mathrm{e}} \rvert$, when ν_{e} is the stoichiometric number of the electron in the electrode reaction.

Name	*Symbol*	*Definition*	*SI unit*	*Notes*
equilibrium electrode potential	E_{eq}	$E_{\mathrm{eq}} = E^{\ominus} - \frac{RT}{zF}\sum_i \nu_i \ln a_i$	V	10–13
formal potential	$E^{\ominus\prime}$	$E_{\mathrm{eq}} = E^{\ominus\prime} - \frac{RT}{zF}\sum_i \nu_i \ln\left(\frac{c_i}{c^{\ominus}}\right)$	V	12, 14
liquid junction potential	E_{j}		V	15
electric current	I	$I = \mathrm{d}Q/\mathrm{d}t$	A	16
electric current density	$j, \boldsymbol{j}$	$j = I/A$	$\mathrm{A\ m^{-2}}$	17
Faradaic current	I_{F}	$I_{\mathrm{F}} = I_{\mathrm{c}} + I_{\mathrm{a}}$	A	18
reduction rate constant	k_{c}	$I_{\mathrm{c}} = -nFAk_{\mathrm{c}}\prod_{\mathrm{B}}(c_{\mathrm{B}}')^{\,n_{\mathrm{B}}}$	(varies)	1, 12, 19
oxidation rate constant	k_{a}	$I_{\mathrm{a}} = nFAk_{\mathrm{a}}\prod_{\mathrm{B}}(c_{\mathrm{B}}')^{\,n_{\mathrm{B}}}$	(varies)	1, 12, 19
overpotential	η, E_{η}	$\eta = E - E_{\mathrm{eq}}$	V	
Tafel slope	b	$b = (\partial E/\partial \ln\lvert I_{\mathrm{F}}\rvert)_{c_i,T,p}$	V	20
mass transfer coefficient	k_{d}	$k_{\mathrm{d,B}} = \lvert\nu_B\rvert\, I_{\mathrm{lim,B}}/nFcA$	$\mathrm{m\ s^{-1}}$	1, 12, 21
electrokinetic potential (ζ-potential)	ζ		V	
electric mobility	$u, (m)$	$u_{\mathrm{B}} = \lvert v_{\mathrm{B}}\rvert/\lvert\boldsymbol{E}\rvert$	$\mathrm{m^2\ V^{-1}\ s^{-1}}$	1, 22
ionic conductivity, molar conductivity of an ion	λ	$\lambda_{\mathrm{B}} = \lvert z_{\mathrm{B}}\rvert F u_{\mathrm{B}}$	$\mathrm{S\ m^2\ mol^{-1}}$	1, 23
molar conductivity	Λ	$\Lambda(\mathrm{A}_{\nu^+}\mathrm{B}_{\nu^-}) = \nu_+\lambda_+ + \nu_-\lambda_-$	$\mathrm{S\ m^2\ mol^{-1}}$	1, 23

(13) This is the Nernst equation. $\sum \nu_i \ln a_i$ refers to the electrode reaction written as a reduction, where a_i are the activities of the species taking part in it; ν_i are the stoichiometric numbers of these species (see p. 45).

(14) It is $E^{\ominus\prime}$ which is calculated in electrochemical experiments when the concentrations of the various species are known, but not their activities.

(15) E_{j} is the Galvani potential difference between two electrolyte solutions in contact.

(16) Q is the charge transferred through the external circuit of the cell.

(17) Formally, the current density is a vector, $\mathrm{d}I = \boldsymbol{j}\cdot\boldsymbol{e}_{\mathrm{n}}\mathrm{d}A$ (see Section 2.3, note 2, p. 24).

(18) I_{F} is the current through the electrode|solution interface, resulting from the charge transfer due to an electrode reaction proceeding as reactants + ne^- $\rightarrow$ products. I_{c} is the *cathodic* partial current due to the reduction reaction, I_{a} is the *anodic* partial current due to the oxidation reaction. By definition I_{c} is negative and I_{a} positive. At the equilibrium potential, $I_{\mathrm{a}} = -I_{\mathrm{c}} = I_0$ (the exchange current) and $j_{\mathrm{a}} = -j_{\mathrm{c}} = j_0$ (the exchange current density). I or j may achieve limiting values, indicated by the subscript lim, in addition to the subscripts F, c, or a.

(19) For a first-order reaction the SI unit is $\mathrm{m\ s^{-1}}$. Here n (or z) is the number of electrons transferred in the electrochemical reaction, c_{B}' is the concentration at the interphase, n_{B} is the order of the reaction with respect to entity B. The formerly used symbols $k_{\mathrm{red}}, k_{\mathrm{f}}$ and $\overrightarrow{k}$ (for k_{c}) and $k_{\mathrm{ox}}, k_{\mathrm{b}}$ and $\overleftarrow{k}$ (for k_{a}) are not recommended.

(20) The Tafel slope b is an experimental quantity from which kinetic information can be derived.

(21) The mass transfer coefficient is the flux density divided by the concentration. For steady-state mass transfer, $k_{\mathrm{d,B}} = D_{\mathrm{B}}/\delta_{\mathrm{B}}$ where δ_{B} is the diffusion layer thickness (which is model dependent) and D_{B} is the diffusion coefficient of entity B. For more information see [22].

(22) v_{B} is the migration velocity of entities B and $\lvert\boldsymbol{E}\rvert$ is the electric field strength within the phase concerned.

(23) The unit $\mathrm{S\ cm^2\ mol^{-1}}$ is often used for molar conductivity. The conductivity is equal to $\kappa = \sum \lambda_i c_i$. It is important to specify the entity to which molar conductivity refers (see Green Book for details [20]). Concerning the name "molar conductivity", see footnote 2 on p. 4.

4.13.1 Sign and notation conventions in electrochemistry

The conventions here are in accordance with the "Stockholm Convention" of 1953 [80].

(i) Electrochemical cells
Electrochemical cells consist of at least two (usually metallic) electron conductors in contact with ionic conductors (electrolytes). Electrochemical cells with current flow can operate either as *galvanic cells*, in which chemical reactions occur spontaneously and chemical energy is converted into electrical energy, or as *electrolytic cells*, in which electrical energy is converted into chemical energy. In both cases part of the energy becomes converted into (positive or negative) heat.

(ii) Electrode
The term electrode means either (i) the electron conductor (usually a metal) connected to the external circuits or (ii) the half cell consisting of one electron conductor and at least one ionic conductor. The latter version has usually been favored.

(iii) Representation of electrochemical cells
Electrochemical cells are represented by diagrams such as those in the following examples:

Examples Pt(s) | H_2(g) | HCl(aq) | AgCl(s) | Ag(s)

Cu(s) | $CuSO_4$(aq) ⁝ $ZnSO_4$(aq) | Zn(s)

Cu(s) | $CuSO_4$(aq) ⁝⁝ KCl(aq, sat) ⁝⁝ $ZnSO_4$(aq) | Zn(s)

A single vertical bar (|) should be used to represent a phase boundary, a dashed vertical bar (⁝) to represent a junction between miscible liquids, and double dashed vertical bars (⁝⁝) to represent a liquid junction, in which the liquid junction potential is assumed to be eliminated.

(iv) Potential difference of an electrochemical cell, anode and cathode
The potential difference of an electrochemical cell is measured between a metallic conductor attached to the right-hand electrode of the cell diagram and an identical metallic conductor attached to the left-hand electrode. Electric potential differences can be measured only between two pieces of material of the same composition. In practice, these are almost always two pieces of copper, attached to the electrodes of the cell.
The terms anode and cathode may only be applied to electrochemical cells through which a net current flows. In a cell at equilibrium the terms plus pole and minus pole are used. An *anode* is an electrode at which the predominating electrochemical reaction is an oxidation; electrons are produced (in a galvanic cell) or extracted (in an electrolytic cell). A *cathode* is an electrode at which the predominating electrochemical reaction is a reduction which consumes the electrons from the anode, and that reach the cathode through the external circuit.

(v) Standard potential of an electrochemical cell reaction
If no current flows through the cell, and all local charge transfer and local chemical equilibria of each electrode reaction are established, the potential difference of the cell is related to the Gibbs energy of the overall cell reaction by the equation

$$\Delta_r G = -zFE_{\text{cell,eq}} = -zF\left(E^{\ominus}{}_{\text{cell,eq}} - \frac{RT}{zF}\sum_i \nu_i \ln a_i\right)$$

assuming that junction potentials are negligible.

(vi) Standard electrode potential (of an electrochemical reaction)
The *standard potential* of an electrochemical reaction, abbreviated as standard electrode potential, is defined as the standard potential of a hypothetical cell, in which the electrode (half-cell) at the left of the cell diagram is the *standard hydrogen electrode* (SHE) and the electrode at the right is the electrode in question. The standard hydrogen electrode consists of a platinum electrode in contact with a solution of H^+ at unit activity and saturated with H_2 gas with a fugacity referred to the standard pressure of 10^5 Pa (see Section 4.11.1 (v), p. 53). For a metallic electrode in equilibrium with solvated ions the cell diagram is

$$\mathrm{Pt} \mid \mathrm{H_2} \mid \mathrm{H^+} \,\vdots\vdots\, \mathrm{M}^{z+} \mid \mathrm{M}$$

and relates to the reaction

$$\mathrm{M}^{z+} + (z/2)\ \mathrm{H_2(g)} = \mathrm{M} + z\ \mathrm{H^+}$$

This diagram may be abbreviated $E(\mathrm{M}^{z+}/\mathrm{M})$. Note that the standard hydrogen electrode as defined is limited to aqueous solutions. See the Green Book for more information (Section 2.13.1 (vi), p. 74 in [20]).

4.14 COLLOID AND SURFACE CHEMISTRY

The recommendations given here are based on more extensive IUPAC recommendations [1.e–1.h] and [85–88]. Catalyst characterization is described in [89] and quantities related to macromolecules in [90].

Name	*Symbol*	*Definition*	*SI unit*	*Notes*
specific surface area	a_s, a, s	$a_\mathrm{s} = A/m$	$\mathrm{m^2\ kg^{-1}}$	1
adsorbed amount of B	${n_\mathrm{B}}^\mathrm{a}$, $n_\mathrm{a}(\mathrm{B})$		mol	2
surface excess amount of B	${n_\mathrm{B}}^\sigma$		mol	3
surface excess concentration of B	$\Gamma_\mathrm{B}, ({\Gamma_\mathrm{B}}^\sigma)$	$\Gamma_\mathrm{B} = {n_\mathrm{B}}^\sigma/A$	$\mathrm{mol\ m^{-2}}$	3
area per molecule	a, σ	$a_\mathrm{B} = A/{N_\mathrm{B}}^\mathrm{a}$	$\mathrm{m^2}$	4
surface coverage	θ	$\theta = {N_\mathrm{B}}^\mathrm{a}/N_\mathrm{m,B}$	1	4
contact angle	θ		rad, 1	
surface tension, interfacial tension	γ, σ	$\gamma = (\partial G/\partial A_\mathrm{s})_{T,p,n_i}$	$\mathrm{N\ m^{-1}}$, $\mathrm{J\ m^{-2}}$	
Debye length of the diffuse layer	L_D	$L_\mathrm{D} = \kappa^{-1}$	m	5
sedimentation coefficient	s	$s = v/a$	s	6
surface pressure	π	$\pi = \gamma^0 - \gamma$	$\mathrm{N\ m^{-1}}$	7

(1) The subscript s designates any surface area where adsorption or deposition may occur. m designates the mass of a solid adsorbent.

(2) The value of ${n_\mathrm{B}}^\mathrm{a}$ depends on the thickness assigned to the surface layer (see also note 1, p. 44).

(3) The values of ${n_\mathrm{B}}^\sigma$ and Γ_B depend on the convention used to define the position of the Gibbs surface. They are given by the excess amount of B or surface concentration of B over values that would apply if each of the two bulk phases were homogeneous right up to the Gibbs dividing surface. See [1.e], and also additional recommendations on this page.

(4) ${N_\mathrm{B}}^\mathrm{a}$ is the number of adsorbed molecules (${N_\mathrm{B}}^\mathrm{a} = N_\mathrm{A}\, {n_\mathrm{B}}^\mathrm{a}$), and ${N_\mathrm{m,B}}^\mathrm{a}$ is the number of adsorbed molecules in a filled monolayer. The definition applies to entities B.

(5) The characteristic Debye length [1.e] and [88] or Debye screening length [88] L_D appears in Gouy-Chapman theory and in the theory of semiconductor space-charge.

(6) In the definition, v is the speed of sedimentation and a is the acceleration of gravity or centrifugation. The symbol for a limiting sedimentation coefficient is $[s]$, for a reduced sedimentation coefficient s°, and for a reduced limiting sedimentation coefficient $[s^\circ]$; see [1.e] for further details.

(7) In the definition, γ^0 is the surface tension of the clean surface and γ that of the covered surface.

The following abbreviations are used in colloid chemistry:

ccc	critical coagulation concentration
cmc	critical micellisation concentration
iep	isoelectric point
pzc	point of zero charge

4.15 TRANSPORT PROPERTIES

The names and symbols recommended here are in agreement with those recommended by IUPAP [4] and ISO [5.f]. Further information on transport phenomena in electrochemical systems can also be found in [81].

Name	*Symbol*	*Definition*	*SI unit*	*Notes*
flux of mass m	q_m	$q_m = \mathrm{d}m/\mathrm{d}t$	kg s^{-1}	1
flux density of mass m	J_m	$J_m = q_m/A$	kg m^{-2} s^{-1}	2, 3
heat flux, thermal power	Φ, P	$\Phi = \mathrm{d}Q/\mathrm{d}t$	W	1
heat flux density	J_q	$J_q = \Phi/A$	W m^{-2}	3
thermal conductance	G	$G = \Phi/\Delta T$	W K^{-1}	
thermal resistance	R	$R = 1/G$	K W^{-1}	
thermal conductivity	λ, k	$\lambda = -J_q/(\mathrm{d}T/\mathrm{d}x)$	W m^{-1} K^{-1}	4
coefficient of heat transfer	$h, (k, K, \alpha)$	$h = J_q/\Delta T$	W m^{-2} K^{-1}	
thermal diffusivity	a	$a = \lambda/\rho c_p$	m^2 s^{-1}	5
diffusion coefficient	D	$D = -J_x\ (\mathrm{d}c/\mathrm{d}x)^{-1}$	m^2 s^{-1}	6
viscosity, shear viscosity	η	$\eta = -\tau_{xz}(\partial v_x/\partial y + \partial v_y/\partial x)^{-1}$	Pa s	7
bulk viscosity	ζ	$\zeta = \dfrac{(\tau_{xx} + \tau_{yy} + \tau_{zz})}{3\,(\boldsymbol{\nabla}\cdot\boldsymbol{v})}$	Pa s	7
thermal diffusion coefficient	D_T	$D_\mathrm{T} = J_x c^{-1}(\mathrm{d}T/\mathrm{d}x)^{-1}$	m^2 K^{-1} s^{-1}	6

(1) A vector quantity, $\boldsymbol{q}_m = (\mathrm{d}m/\mathrm{d}t)\,\boldsymbol{e}$, where $\boldsymbol{e}$ is the unit vector in the direction of the flow, can be defined, sometimes called the "flow rate" of m. Similar definitions apply to any extensive quantity.
(2) The flux density of molecules, J_N, determines either the rate at which the surface would be covered if each molecule stuck, or the rate of effusion through a hole in the surface. In studying the exposure of a surface to a gas, surface scientists find it useful to use the product of pressure and time as a measure of the exposure, since this product is proportional to the number flux density and time. The unit langmuir (symbol L) corresponds to the exposure of a surface to a gas at 10^{-6} Torr for 1 s.
(3) In previous editions of the Green Book [2.a,2.b], the term "flux" was used for what is now called "flux density", in agreement with IUPAP [4]. The term "density" in the name of an intensive physical quantity usually implies "extensive quantity divided by volume" for scalar quantities but "extensive quantity divided by area" for vector quantities denoting flow or flux. A is an area. In the old literature a statement like "the heat flux is 10 W m^{-2}" is to be interpreted as "the heat flux density is 10 W m^{-2}" in modern language. Similarly, the term "flow" was used for what is now "flux", so that "the heat flow is 100 W" in old literature means today "the heat flux is 100 W".
(4) J_q is the heat flux density in direction of x.
(5) ρ is the density and c_p the specific heat capacity at constant pressure.
(6) J_x is the amount flux density component in direction of x and c is the amount concentration.
(7) See also Section 4.2, p. 23; τ is the shear stress tensor. $v_{x,y,z}$ are the Cartesian flow velocity components.

4.15.1 Transport characteristic numbers: Quantities of dimension one

The following symbols are used in the definitions: density (ρ), speed (v), length (l), viscosity (η), acceleration of gravity (g), cubic expansion coefficient (α), thermodynamic temperature (T), surface tension (γ), speed of sound (c), thermal diffusivity (a), coefficient of heat transfer (h), specific heat capacity at constant pressure (c_p), diffusion coefficient (D), mass transfer coefficient (k_d).

Name	*Symbol*	*Definition*	*Notes*
Reynolds number	Re	$Re = \rho v l/\eta$	
Grashof number	Gr	$Gr = l^3 g \alpha \Delta T \rho^2/\eta^2$	
Weber number	We	$We = \rho v^2 l/\gamma$	
Mach number	Ma	$Ma = v/c$	
Péclet number	Pe	$Pe = vl/a$	
Stanton number	St	$St = h/\rho v c_p$	
Nusselt number for mass transfer	Nu^*	$Nu^* = k_\mathrm{d} l/D$	1
Prandtl number	Pr	$Pr = \eta/\rho a$	
Schmidt number	Sc	$Sc = \eta/\rho D$	

(1) This quantity applies to the transport of matter in binary mixtures [61]. The name Sherwood number and symbol Sh have been widely used for this quantity.

5 UNCERTAINTY

The tables present a summary of basic principles based on the GUM [8] approach to estimating and reporting uncertainty budgets for experimental measurement results.

5.1 REPORTING UNCERTAINTY FOR A SINGLE MEASURED QUANTITY

Uncertainties for a measured quantity X can be reported in several ways by giving the expected best estimate for the quantity with an uncertainty which corresponds to some probable range for the true quantity. We use the notation X for measured quantities, x for measured values, and u_c for the combined standard uncertainty. All definitions given in this chapter apply to uncorrelated values. When reporting the uncertainty of a measured value, a note must be given explaining what the uncertainty means.

Example 1 $m_s = 100.021\,47(35)$ g, where the number in parentheses denotes the combined standard uncertainty $u_c = 0.35$ mg and is assumed to apply to the least significant digits.

If the measured values x given by some real number multiplied by some unit are approximately normally distributed, the unknown value of the quantity X is estimated to lie within the interval defined by U with a level of confidence of 95 %.

Example 2 $m_s = (100.021\,47 \pm 0.000\,70)$ g, where the number following the symbol ± is the numerical value of an expanded uncertainty $U = k\,u_c$, with U determined from a combined standard uncertainty $u_c = 0.35$ mg and a coverage factor of $k = 2$.

Name	*Symbol*	*Definition*	*Notes*
Statistics			
number of measurements	N		
mean	$\bar{x}$	$\bar{x} = \frac{1}{N}\sum_{i=1}^{N} x_i$	
variance	$s^2(x)$	$s^2(x) = \frac{1}{N-1}\sum_{i=1}^{N} (x_i - \bar{x})^2$	1
standard deviation	$s(x)$	$s(x) = \lvert\sqrt{s^2(x)}\rvert$	
standard deviation of the mean	$s(\bar{x})$	$s(\bar{x}) = s(x)/\sqrt{N}$	
Uncertainties			
standard uncertainty of x_i	$u(x_i)$	Estimated by Type A or B approaches	2
Type A		Type A - statistical analysis	3
Type B evaluation		Type B - prior, background or other information	4
relative standard uncertainty of x	$u_r(x_i)$	$u_r(x_i) = u(x_i)/\lvert x_i\rvert \quad (x_i \neq 0)$	
expanded uncertainty	U	$U = k\,u_c(y)$	
coverage factor	k		

(1) This is an unbiased estimate and takes into account the removal of one degree of freedom because the spread is measured about the mean.

(2) An uncertainty evaluation generally consists of several components:
Type A: Those that are evaluated by statistical methods
Type B: Those that are evaluated by other means

(3) Examples of type A evaluations are calculating the standard deviation of the mean of a series of independent observations using the method of least squares to fit a function to data in order to estimate

Some commercial, industrial, and regulatory applications require a measure of uncertainty that defines an interval about a measured value y within which the value of the measured quantity Y can be confidently asserted to lie. In these cases the *expanded uncertainty* U is used, and is obtained by multiplying the combined standard uncertainty, $u_\mathrm{c}(y)$, by a *coverage factor* k. Thus $U = k\, u_\mathrm{c}(y)$ and it is confidently believed that Y is greater than or equal to $y - U$, and is less than or equal to $y + U$, which is commonly written as $Y = y \pm U$. Typically, k is in the range of two to three. When the normal distribution applies and u_c is a reliable estimate of the standard uncertainty of y, $U = 2u_\mathrm{c}$ (i.e., $k = 2$) defines an interval having a level of confidence of approximately 95 %, and $U = 3u_\mathrm{c}$ (i.e., $k = 3$) defines an interval having a level of confidence greater than 99 %. This interval can be specified as $(y - U) \leq Y \leq (y + U)$.

Example 3 $100.020\,77\ \mathrm{g} \leq m_\mathrm{s} \leq 100.022\,17\ \mathrm{g}$ or $m_\mathrm{s} = (100.021\,47 \pm 0.000\,70)\ \mathrm{g}$
where 0.7 mg is the expanded uncertainty with a coverage factor 2.

Specifying such an interval implies providing the confidence as the estimated probability with which the true value of the quantity is expected to be in the given range, e.g. 60 %, 90 % or 95 %.
In the frequently encountered situation, where a confidence interval has to be estimated for the value of a quantity Y from the average of a limited number N of observations (the 'sample'), one uses Student's t-distribution. The confidence intervals for y_t are then to be calculated in practice to a good approximation by

$$\bar{y} - \frac{k_t(p)}{\sqrt{N}} s_y \leq y_t \leq \bar{y} + \frac{k_t(p)}{\sqrt{N}} s_y$$

Here, $k_t(p)$ is the factor to be used for the level of confidence p to be specified in percent (i.e. 90 %, 95 % etc.), with the sample mean $\bar{y}$ and the sample standard deviation s_y.

$$\bar{y} = \frac{1}{N}\sum_{i=1}^{N} y_i \qquad s_y = \left(\frac{1}{N-1}\sum_{i=1}^{N}\left(y_i - \bar{y}\right)^2\right)^{1/2} \qquad \delta y_\mathrm{rms} = \left(\frac{1}{N}\sum_{i=1}^{N}\left(y_i - \bar{y}\right)^2\right)^{1/2}$$

The values of $k_t(p)$ are obtained from the integration of the t-distribution and are tabulated below for some values of N and the most common used levels of confidence p (see also [8]). In the limit of a large number of measurements ($N \to \infty$) one approaches the results from a normal Gaussian distribution. Sometimes, the spread of values in a sample is characterized by the 'root mean square deviation', δy_rms.

N	2	3	4	5	6	7	8	9	10	∞
$k_t(p = 60\,\%)$	1.38	1.06	0.98	0.94	0.92	0.91	0.90	0.89	0.88	0.84
$k_t(p = 68\,\%)$	1.82	1.31	1.19	1.13	1.10	1.08	1.07	1.06	1.05	0.99
$k_t(p = 90\,\%)$	6.31	2.92	2.35	2.13	2.02	1.94	1.89	1.86	1.83	1.64
$k_t(p = 95\,\%)$	12.71	4.30	3.18	2.78	2.57	2.45	2.36	2.31	2.26	1.96
$k_t(p = 99\,\%)$	63.66	9.92	5.84	4.60	4.03	3.71	3.50	3.36	3.25	2.58
$k_t(p = 99.8\,\%)$	318.31	22.33	10.21	7.17	5.89	5.21	4.79	4.50	4.30	3.09

Example 4 An integrated absorption band intensity has been determined as integrated absorption cross section (see Section 4.7.1) from five independent measurements with an average value $\bar{G} = 2.031\ \mathrm{pm}^2$ and a sample standard deviation $s_G = 0.052\ \mathrm{pm}^2$. One would calculate the expanded uncertainty as $2.13/\sqrt{5} \cdot 0.052\,\mathrm{pm}^2 \approx 0.050\ \mathrm{pm}^2$ and report $G = (2.031 \pm 0.050)\ \mathrm{pm}^2$ at 90 % confidence or $G = (2.031 \pm 0.065)\ \mathrm{pm}^2$ at 95 % confidence; the level of confidence must be specified.

(Notes continued)
the parameters of the function and their standard deviations.
(4) Type B may include previous measurement data, general knowledge of the behavior and properties of materials and instruments, manufacturer's specifications, data provided in calibration and other reports.

6 REFERENCES

The primary sources are given in [1a] to [9] and the secondary sources begin with [10]. A more complete list of references can be found in [2c].

6.1 PRIMARY SOURCES

[1] Manual of Symbols and Terminology for Physicochemical Quantities and Units
(a) 1st ed., M. L. McGlashan. *Pure Appl. Chem.*, 21:1–38, 1970.
(b) 2nd ed., M. A. Paul. Butterworths, London, 1975.
(c) 3rd ed., D. H. Whiffen. *Pure Appl. Chem.*, 51:1–36, 1979.
(d) D. H. Whiffen. Appendix I–Definitions of Activities and Related Quantities. *Pure Appl. Chem.*, 51:37–41, 1979.
(e) D. H. Everett. Appendix II–Definitions, Terminology and Symbols in Colloid and Surface Chemistry, Part I. *Pure Appl. Chem.*, 31:577–638, 1972.
(f) J. Lyklema and H. van Olphen. Appendix II–Definitions, Terminology and Symbols in Colloid and Surface Chemistry, Part 1.13: Selected Definitions, Terminology and Symbols for Rheological Properties. *Pure Appl. Chem.*, 51:1213–1218, 1979.
(g) M. Kerker and J. P. Kratochvil. Appendix II–Definitions, Terminology and Symbols in Colloid and Surface Chemistry, Part 1.14: Light Scattering. *Pure Appl. Chem.*, 55:931–941, 1983.
(h) R. L. Burwell Jr. Part II: Heterogeneous Catalysis. *Pure Appl. Chem.*, 46:71–90, 1976.
(i) R. Parsons. Appendix III–Electrochemical Nomenclature. *Pure Appl. Chem.*, 37:499–516, 1974.
(j) J. D. Cox. Appendix IV–Notation for States and Processes, Significance of the Word "Standard" in Chemical Thermodynamics, and Remarks on Commonly Tabulated Forms of Thermodynamic Functions. *Pure Appl. Chem.*, 54:1239–1250, 1982.
(k) K. J. Laidler. Appendix V–Symbolism and Terminology in Chemical Kinetics. *Pure Appl. Chem.*, 53:753–771, 1981.

[2] (a) I. Mills, T. Cvitaš, K. Homann, N. Kallay and K. Kuchitsu. *Quantities, Units and Symbols in Physical Chemistry.* Blackwell Science, Oxford, 1st edition, 1988.
(b) I. Mills, T. Cvitaš, K. Homann, N. Kallay and K. Kuchitsu. *Quantities, Units and Symbols in Physical Chemistry.* Blackwell Science, Oxford, 2nd edition, 1993.
(c) E. R. Cohen, T. Cvitaš, J. G. Frey, B. Holmström, K. Kuchitsu, R. Marquardt, I. Mills, F. Pavese, M. Quack, J. Stohner, H. L. Strauss, M. Takami, A. J. Thor. *Quantities, Units and Symbols in Physical Chemistry.* IUPAC & The Royal Society of Chemistry, Cambridge, 3rd edition 2007, 3rd printing, 2011.
(d) *Nomenklaturniye Pravila IUPAC po Khimii, Vol. 6, Fizicheskaya Khimiya,* Nacionalnii Komitet Sovetskih Khimikov, Moscow, 1988.
(e) M. Riedel. *A Fizikai-kémiai Definiciók és Jelölések.* Tankönyvkiadó, Budapest, 1990.
(f) K. Kuchitsu. *Butsurikagaku de Mochiirareru Ryo, Tan-i, Kigo.* Kodansha, Tokyo, 1991.
(g) K.-H. Homann, M. Hausmann. *Größen, Einheiten und Symbole in der Physikalischen Chemie.* VCH, Weinheim, 1996.
(h) D. I. Marchidan. *Mărimi, Unităţi şi Simboluri în Chimia Fizică.* Editura Academiei Române, Bucharest, 1996.
(i) A. P. Masiá, J. M. Guil, J. E. Herrero, A. R. Paniego. *Magnitudes, Unidades y Símbolos en Química Física.* Fundació Ramón Areces & Real Sociedad Española de Química, Madrid, 1999.
(j) J. M. Costa. *Magnituds, Unitats i Símbols en Química Física.* Institut d'Estudis Catalans, Barcelona, 2004.

(k) E. R. Cohen, T. Cvitaš, J. G. Frey, B. Holmström, K. Kuchitsu, R. Marquardt, I. Mills, F. Pavese, M. Quack, J. Stohner, H. L. Strauss, M. Takami, A. J. Thor.
Grandeurs, unités et symboles de la chimie physique. De Boeck, Bruxelles, 2012
Trad. de la 3ème édition anglaise par R. Marquardt, M. Mottet, F. Rouquérol, J. Toullec.

(l) R. C. Rocha-Filho, R. Fausto.
Grandezas, Unidades e Símbolos em Físico-Química. EditSBQ, 2018.
(available online `http://www.sbq.org.br/livroverde/`)

(m) E. R. Cohen, T. Cvitaš, J. G. Frey, B. Holmström, K. Kuchitsu, R. Marquardt, I. Mills, F. Pavese, M. Quack, J. Stohner, H. L. Strauss, M. Takami, A. J. Thor.
Quantities, Units and Symbols in Physical Chemistry,
translated in Japanese by the Chemical Society of Japan,
and the National Metrology Institute, National Institute of Advanced Science and Technology, Kodansha Scientific, Tokyo, 2009.

[3] Bureau International des Poids et Mesures. *Le Système International d'Unités (SI).* 9th French and English Edition, BIPM, Sèvres, 2019
(available online `https://www.bipm.org/en/publications/si-brochure/`).

[4] E. R. Cohen and P. Giacomo. *Symbols, Units, Nomenclature and Fundamental Constants in Physics. 1987 Revision*, Document I.U.P.A.P.–25 (IUPAP-SUNAMCO 87–1)
also published in: *Physica*, 146A:1–67, 1987.

[5] International Standards, ISO
International Organization for Standardization, Geneva, Switzerland

(a)	ISO 80000–1:	2009,	Quantities and units – Part 1:	General
	Technical Corrigendum 1: 2011			
(b)	ISO 80000–2:	2019,	Quantities and units – Part 2:	Mathematics
(c)	ISO 80000–5:	2019,	Quantities and units – Part 5:	Thermodynamics
(d)	ISO 80000–7:	2019,	Quantities and units – Part 7:	Light and radiation
(e)	ISO 80000–9:	2019,	Quantities and units – Part 9:	Physical chemistry and molecular physics
(f)	ISO 80000–11:	2019,	Quantities and units – Part 11:	Characteristic numbers

[6] ISO 1000: 1992, SI Units and recommendations for the use of their multiples and of certain other units
Amendment 1: 1998

[7] International Standard, IEC International Electrotechnical Commission, Geneva.
IEC 60027–2, 2005.

[8] Evaluation of measurement data – Guide to the expression of uncertainty in measurement (GUM), Joint Committee for Guides in Metrology JCGM 100: 2008
(available online `https://www.bipm.org/en/publications/guides/gum.html`).

[9] International vocabulary of metrology – Basic and general concepts and associated Terms (VIM), 3rd edition, 2012
(available online `https://www.bipm.org/en/publications/guides/vim.html`).

6.2 SECONDARY SOURCES

[10] M. Quack. Commission I.1 at the IUPAC General Assembly 1995. Summary Minutes. *Chem. Int.*, 20:12, 1998.

[11] E. A. Guggenheim. Units and Dimensions. *Phil. Mag.*, 33:479–496, 1942.

[12] J. de Boer. On the History of Quantity Calculus and the International System. *Metrologia*, 31:405–429, 1994/95.

[13] I. M. Mills. The Language of Science. *Metrologia*, 34:101–109, 1997.

[14] A. P. Raposo. The Algebraic Structure of Quantity Calculus. *Measurement Science review*, 18:147–157, 2018.

[15] J. C. Rigg, S. S. Brown, R. Dybkaer, and H. Olesen, editors. *Compendium of Terminology and Nomenclature of Properties in Clinical Laboratory Sciences.* Blackwell, Oxford, 1995.

[16] J. Rigaudy and S. P. Klesney. *IUPAC Nomenclature of Organic Chemistry, Sections A, B, C, D, E, F, and H.* Pergamon Press, Oxford, 4th edition, 1979.

[17] E. Tiesinga, P. J. Mohr, D. B. Newell, and B. N. Taylor. CODATA recommended values of the fundamental physical constants: 2018. *J. Phys. Chem. Ref. Data*, 50:033105, 2021.

[18] D. B. Newell, F. Cabiati, J. Fischer, K. Fujii, S. G. Karshenboim, H. S. Margolis, E. de Mirandés, P. J. Mohr, F. Nez, K. Pachucki, T. J. Quinn, B. N. Taylor, M. Wang, B. M. Wood, and Z. Zhang. The CODATA 2017 values of h, e, k, and N_A for the revision of the SI. *Metrologia*, 55:L13–L16, 2018; available online `https://pml.nist.gov/cuu/Constants/`.

[19] N. E. Holden, M. L. Bonardi, P. De Bièvre, P. R. Renne, and I. M. Villa. IUPAC-IUGS common definition and convention on the use of the year as a derived unit of time (IUPAC Recommendations 2011). *Pure and Applied Chemistry*, 83(5):1159–1162, 2011.

[20] E. R. Cohen, T. Cvitaš, J. Frey, B. Holmström, K. Kuchitsu, R. Marquardt, I. Mills, F. Pavese, M. Quack, J. Stohner, H. L. Strauss, M. Takami, and A. J. Thor. *Quantities, Units and Symbols in Physical Chemistry.* IUPAC & The Royal Society of Chemistry, Cambridge, 3rd Printing, 3rd Edition (2011) and 4th Edition in preparation.

[21] A. D. McNaught and A. Wilkinson. *Compendium of Chemical Terminology – The Gold Book*, 2nd edition. Blackwell, Oxford, 1997.

[22] P. Müller. Glossary of Terms Used in Physical Organic Chemistry. *Pure Appl. Chem.*, 66:1077–1184, 1994.

[23] A. D. Jenkins, P. Kratochvíl, R. F. T. Stepto, and U. W. Suter. Glossary of Basic Terms in Polymer Science. *Pure Appl. Chem.*, 68:2287–2311, 1996.

[24] R. D. Brown, J. E. Boggs, R. Hilderbrandt, K. Lim, I. M. Mills, E. Nikitin, and M. H. Palmer. Acronyms Used in Theoretical Chemistry. *Pure Appl. Chem.*, 68:387–456, 1996.

[25] J. Mullay. Estimation of Atomic and Group Electronegativities. *Struct. Bonding (Berlin)*, 66:1–25, 1987.

[26] J. M. Brown, R. J. Buenker, A. Carrington, C. di Lauro, R. N. Dixon, R. W. Field, J. T. Hougen, W. Hüttner, K. Kuchitsu, M. Mehring, A. J. Merer, T. A. Miller, M. Quack, D. A. Ramsey, L. Veseth, and R. N. Zare. Remarks on the Signs of g-Factors in Atomic and Molecular Zeeman Spectroscopy. *Mol. Phys.*, 98:1597–1601, 2000.

[27] F. A. Jenkins. Report of Subcommittee f (Notation for the Spectra of Diatomic Molecules). *J. Opt. Soc. Amer.*, 43:425–426, 1953.

[28] R. S. Mulliken. Report on Notation for the Spectra of Polyatomic Molecules. *J. Chem. Phys.*, 23:1997–2011, 1955. (Erratum) *J. Chem. Phys.*, 24:1118, 1956.

[29] G. Herzberg. *Molecular Spectra and Molecular Structure Vol. II. Infrared and Raman Spectra of Polyatomic Molecules.* Van Nostrand, Princeton, 1946.

[30] G. Herzberg. *Molecular Spectra and Molecular Structure Vol. I. Spectra of Diatomic Molecules.* Van Nostrand, Princeton, 1950.

[31] G. Herzberg. *Molecular Spectra and Molecular Structure Vol. III. Electronic Spectra and Electronic Structure of Polyatomic Molecules.* Van Nostrand, Princeton, 1966.

[32] M. Quack and F. Merkt, editors. *Handbook of High-Resolution Spectroscopy*, volume 1–3. Wiley, Chichester, New York, 2011.

[33] E. D. Becker. Recommendations for the Presentation of Infrared Absorption Spectra in Data Collections – A. Condensed Phases. *Pure Appl. Chem.*, 50:231–236, 1978.

[34] E. D. Becker, J. R. Durig, W. C. Harris, and G. J. Rosasco. Presentation of Raman Spectra in Data Collections. *Pure Appl. Chem.*, 53:1879–1885, 1981.

[35] Physical Chemistry Division, Commission on Molecular Structure and Spectroscopy. Recommendations for the Presentation of NMR Data for Publication in Chemical Journals. *Pure Appl. Chem.*, 29:625–628, 1972.

[36] Physical Chemistry Division, Commission on Molecular Structure and Spectroscopy. Presentation of NMR Data for Publication in Chemical Journals – B. Conventions Relating to Spectra from Nuclei other than Protons. *Pure Appl. Chem.*, 45:217–219, 1976.

[37] Physical Chemistry Division, Commission on Molecular Structure and Spectroscopy. Nomenclature and Spectral Presentation in Electron Spectroscopy Resulting from Excitation by Photons. *Pure Appl. Chem.*, 45:221–224, 1976.

[38] Physical Chemistry Division, Commission on Molecular Structure and Spectroscopy. Nomenclature and Conventions for Reporting Mössbauer Spectroscopic Data. *Pure Appl. Chem.*, 45:211–216, 1976.

[39] J. H. Beynon. Recommendations for Symbolism and Nomenclature for Mass Spectroscopy. *Pure Appl. Chem.*, 50:65–73, 1978.

[40] C. J. H. Schutte, J. E. Bertie, P. R. Bunker, J. T. Hougen, I. M. Mills, J. K. G. Watson, and B. P. Winnewisser. Notations and Conventions in Molecular Spectroscopy: Part 1. General Spectroscopic Notation. *Pure Appl. Chem.*, 69:1633–1639, 1997.

[41] C. J. H. Schutte, J. E. Bertie, P. R. Bunker, J. T. Hougen, I. M. Mills, J. K. G. Watson, and B. P. Winnewisser. Notations and Conventions in Molecular Spectroscopy: Part 2. Symmetry Notation. *Pure Appl. Chem.*, 69:1641–1649, 1997.

[42] P. R. Bunker, C. J. H. Schutte, J. T. Hougen, I. M. Mills, J. K. G. Watson, and B. P. Winnewisser. Notations and Conventions in Molecular Spectroscopy: Part 3. Permutation and Permutation-Inversion Symmetry Notation. *Pure Appl. Chem.*, 69:1651–1657, 1997.

[43] R. K. Harris, E. D. Becker, S. M. Cabral de Menezes, R. Goodfellow, and P. Granger. Parameters and Symbols for Use in Nuclear Magnetic Resonance. *Pure Appl. Chem.*, 69:2489–2495, 1997.

[44] J. L. Markley, A. Bax, Y. Arata, C. W. Hilbers, R. Kaptein, B. D. Sykes, P. E. Wright, and K. Wüthrich. Recommendations for the Presentation of NMR Structures of Proteins and Nucleic Acids. *Pure Appl. Chem.*, 70:117–142, 1998.

[45] J. K. G. Watson. Aspects of Quartic and Sextic Centrifugal Effects on Rotational Energy Levels. In J. R. Durig, editor, *Vibrational Spectra and Structure, Vol. 6*, pages 1–89, Amsterdam, 1977. Elsevier.

[46] R. K. Harris, E. D. Becker, S. M. Cabral de Menezes, R. Goodfellow, and P. Granger. NMR Nomenclature. Nuclear Spin Properties and Conventions for Chemical Shifts. *Pure Appl. Chem.*, 73:1795–1818, 2001.

[47] P. R. Bunker and P. Jensen. *Molecular Symmetry and Spectroscopy,* 2nd edition. NRC Research Press, Ottawa, 1998.

[48] H. C. Longuet-Higgins. The Symmetry Groups of Non-Rigid Molecules. *Mol. Phys.*, 6:445–460, 1963.

[49] M. Quack. Detailed Symmetry Selection Rules for Reactive Collisions. *Mol. Phys.*, 34:477–504, 1977.

[50] I. M. Mills and M. Quack. Introduction to 'The Symmetry Groups of Non-Rigid Molecules' by H. C. Longuet-Higgins. *Mol. Phys.*, 100:9–10, 2002.

[51] J. C. D. Brand, J. H. Callomon, K. K. Innes, J. Jortner, S. Leach, D. H. Levy, A. J. Merer, I. M. Mills, C. B. Moore, C. S. Parmenter, D. A. Ramsay, K. N. Rao, E. W. Schlag, J. K. G. Watson, and R. N. Zare. The Vibrational Numbering of Bands in the Spectra of Polyatomic Molecules. *J. Mol. Spectrosc.*, 99:482–483, 1983.

[52] M. Terazima, N. Hirota, S. E. Braslavsky, A. Mandelis, S. E. Bialkowski, G. J. Diebold, R. J. D. Miller, D. Fournier, R. A. Palmer, and A. Tam. Quantities, Terminology, and Symbols in Photothermal and Related Spectroscopies. *Pure Appl. Chem.*, 76:1083–1118, 2004.

[53] N. Sheppard, H. A. Willis, and J. C. Rigg. Names, Symbols, Definitions and Units of Quantities in Optical Spectroscopy. *Pure Appl. Chem.*, 57:105–120, 1985.

[54] V. A. Fassel. Nomenclature, Symbols, Units and their Usage in Spectrochemical Analysis. I. General Atomic Emission Spectroscopy. *Pure Appl. Chem.*, 30:651–679, 1972.

[55] W. H. Melhuish. Nomenclature, Symbols, Units and their Usage in Spectrochemical Analysis. VI: Molecular Luminescence Spectroscopy. *Pure Appl. Chem.*, 56:231–245, 1984.

[56] J. W. Verhoeven. Glossary of Terms Used in Photochemistry. *Pure Appl. Chem.*, 68:2223–2286, 1996.

[57] A. A. Lamola and M. S. Wrighton. Recommended Standards for Reporting Photochemical Data. *Pure Appl. Chem.*, 56:939–944, 1984.

[58] N. D. Mermin and N. W. Ashcroft. *Solid State Physics.* Holt-Saunders International Editions, New York, 1976.

[59] T. Hahn, editor. *International Tables for Crystallography, Vol. A: Space-Group Symmetry.* Reidel, Dordrecht, 2nd edition, 1983.

[60] S. Trasatti and R. Parsons. Interphases in Systems of Conducting Phases. *Pure Appl. Chem.*, 58:437–454, 1986.

[61] T. Cvitaš. Quantities Describing Compositions of Mixtures. *Metrologia*, 33:35–39, 1996.

[62] M. B. Ewing, T. H. Lilley, G. M. Olofsson, M. T. Rätzsch, and G. Somsen. Standard Quantities in Chemical Thermodynamics. Fugacities, Activities, and Equilibrium Constants for Pure and Mixed Phases. *Pure Appl. Chem.*, 66:533–552, 1994.

[63] W. H. Powell. Revised Nomenclature for Radicals, Ions, Radical Ions and Related Species. *Pure Appl. Chem.*, 65:1357–1455, 1993.

[64] W. H. Koppenol. Names for Inorganic Radicals. *Pure Appl. Chem.*, 72:437–446, 2000.

[65] N. G. Connelly, T. Damhus, R. M. Hartshorn, and A. T. Hutton, editors. *Nomenclature of Inorganic Chemistry – IUPAC Recommendations 2005.* The Royal Society of Chemistry, Cambridge, 2005.

[66] E. S. Domalski. Selected Values of Heats of Combustion and Heats of Formation of Organic Compounds Containing the Elements C, H, N, O, P, and S. *J. Phys. Chem. Ref. Data*, 1:221–277, 1972.

[67] R. D. Freeman. Conversion of Standard (1 atm) Thermodynamic Data to the New Standard-State Pressure, 1 bar (10^5 Pa). *Bull. Chem. Thermodyn.*, 25:523–530, 1982.

[68] R. D. Freeman. Conversion of Standard (1 atm) Thermodynamic Data to the New Standard-State Pressure, 1 bar (10^5 Pa). *J. Chem. Eng. Data*, 29:105–111, 1984.

[69] R. D. Freeman. Conversion of Standard Thermodynamic Data to the New Standard-State Pressure. *J. Chem. Educ.*, 62:681–686, 1985.

[70] I. Tinoco Jr., K. Sauer, and J. C. Wang. *Physical Chemistry,* 4th edition. Prentice-Hall, New Jersey, 2001.

[71] R. A. Alberty. Recommendations for Nomenclature and Tables in Biochemical Thermodynamics. *Pure Appl. Chem.*, 66:1641–1666, 1994.

[72] D. D. Wagman, W. H. Evans, V. B. Parker, R. H. Schumm, I. Halow, S. M. Bailey, K. L. Churney, and R. L. Nuttall. The NBS Tables of Chemical Thermodynamic Properties. *J. Phys. Chem. Ref. Data, Vol. 11, Suppl. 2*, 1982.

[73] M. W. Chase Jr., C. A. Davies, J. R. Downey Jr., D. J. Frurip, R. A. McDonald, and A. N. Syverud. JANAF Thermochemical Tables, 3rd edition. *J. Phys. Chem. Ref. Data, Vol. 14, Suppl. 1*, 1985.

[74] V. P. Glushko, editor. *Termodinamicheskie Svoistva Individualnykh Veshchestv, Vols. 1-4.* Nauka, Moscow, 1978-85.

[75] I. Barin, editor. *Thermochemical Data of Pure Substances,* 3rd edition. VCH, Weinheim, 1995.

[76] M. W. Chase Jr. NIST JANAF Thermochemical Tables, 4th edition. *J. Phys. Chem. Ref. Data, Monograph 9, Suppl.*, 1998.

[77] CODATA Task Group on Data for Chemical Kinetics. The Presentation of Chemical Kinetics Data in the Primary Literature. *CODATA Bull.*, 13:1–7, 1974.

[78] K. J. Laidler. A Glossary of Terms Used in Chemical Kinetics, Including Reaction Dynamics. *Pure Appl. Chem.*, 68:149–192, 1996.

[79] M. L. Goldberger and K. M. Watson. *Collision Theory.* Krieger, Huntington (NY), 1975.

[80] P. van Rysselberghe. Internationales Komitee für elektrochemische Thermodynamik und Kinetik CITCE. Bericht der Kommission für elektrochemische Nomenklatur und Definitionen. *Z. Elektrochem.*, 58:530–535, 1954.

[81] R. Parsons. Electrode Reaction Orders, Transfer Coefficients and Rate Constants. Amplification of Definitions and Recommendations for Publication of Parameters. *Pure Appl. Chem.*, 52:233–240, 1980.

[82] N. Ibl. Nomenclature for Transport Phenomena in Electrolytic Systems. *Pure Appl. Chem.*, 53:1827–1840, 1981.

[83] S. Trasatti. The Absolute Electrode Potential: An Explanatory Note. *Pure Appl. Chem.*, 58:955–966, 1986.

[84] A. J. Bard, R. Parsons, and J. Jordan, editors. *Standard Potentials in Aqueous Solutions.* Marcel Dekker Inc., New York, 1985.

[85] K. S. W. Sing, D. H. Everett, R. A. W. Haul, L. Moscou, R. A. Pierotti, J. Rouquérol, and T. Siemieniewska. Reporting Physisorption Data for Gas/Solid Systems with Special Reference to the Determination of Surface Area and Porosity. *Pure Appl. Chem.*, 57:603–619, 1985.

[86] L. Ter-Minassian-Saraga. Reporting Experimental Pressure-Area Data with Film Balances. *Pure Appl. Chem.*, 57:621–632, 1985.

[87] D. H. Everett. Reporting Data on Adsorption from Solution at the Solid/Solution Interface. *Pure Appl. Chem.*, 58:967–984, 1986.

[88] L. Ter-Minassian-Saraga. Thin Films Including Layers: Terminology in Relation to their Preparation and Characterization. *Pure Appl. Chem.*, 66:1667–1738, 1994.

[89] J. Haber. Manual on Catalyst Characterization. *Pure Appl. Chem.*, 63:1227–1246, 1991.

[90] W. V. Metanomski. *Compendium of Macromolecular Nomenclature.* Blackwell, Oxford, 1991.

7 GREEK ALPHABET

Roman	Italics	*Name*	*Pronounciation and Latin Equivalent*	*Notes*
Α, α	A, α	alpha	A	
Β, β	B, β	beta	B	
Γ, γ	Γ, γ	gamma	G	
Δ, δ	Δ, δ	delta	D	
Ε, ε	E, ε, ϵ	epsilon	E	
Ζ, ζ	Z, ζ	zeta	Z	
Η, η	H, η	eta	Ae, Ä, Ee	1
Θ, ϑ, θ	Θ, ϑ, θ	theta	Th	2
Ι, ι	I, ι	iota	I	
Κ, ϰ, κ	K, $\varkappa$, κ	kappa	K	2
Λ, λ	Λ, λ	lambda	L	
Μ, μ	M, μ	mu, (my)	M	
Ν, ν	N, ν	nu, (ny)	N	
Ξ, ξ	Ξ, ξ	xi	X	
Ο, ο	O, o	omikron	O	
Π, π	Π, π	pi	P	
Ρ, ρ	P, ρ	rho	R	
Σ, σ, ς	Σ, σ, ς	sigma	S	2, 3
Τ, τ	T, τ	tau	T	
Υ, υ	Y, υ	upsilon, ypsilon	U, Y	
Φ, φ, ϕ	Φ, φ, ϕ	phi	Ph	2
Χ, χ	X, χ	chi	Ch	
Ψ, ψ	Ψ, ψ	psi	Ps	
Ω, ω	Ω, ω	omega	Oo	4

(1) For the Latin equivalent Ae is to be pronounced as the German Ä. The modern Greek pronounciation of the letter η is like E, long ee as in cheese, or short i as in lips. The Latin equivalent is also often called "long E".
(2) For the lower case letters epsilon, theta, kappa, sigma and phi there are two variants in each case. For instance, the second variant of the lower case theta is sometimes called "vartheta" in printing.
(3) The second variant for lower case sigma is used in Greek only at the end of the word.
(4) In contrast to omikron (short o) the letter omega is pronounced like a long o.

8 INDEX OF SYMBOLS

This index lists symbols for physical quantities, units, some selected mathematical operators, states of aggregation, processes, and particles. Qualifying subscripts and superscripts, etc., are generally omitted from this index, so that for example E_{p} for potential energy, and E_{ea} for electron affinity are both indexed under E for energy. The Latin alphabet is indexed ahead of the Greek alphabet, lower case letters ahead of upper case, bold ahead of italic, ahead of roman, and single letter symbols ahead of multiletter ones. When more than one page reference is given, bold print is used to indicate the general (defining) reference. Numerical entries for the corresponding quantities are underlined. The acronym 'ifc' for a page number refers to the inside front cover.

a acceleration, 17, **22**
a fundamental translation vector, **41**
a* reciprocal lattice vector, **41**
a absorption coefficient, 4, **38**
a activity, **49**, 50, 51, 59, 60, 61
a area per molecule, **63**
a hyperfine coupling constant, 31
a specific surface area, 63
a thermal diffusivity, 64
a unit cell length, 41
a van der Waals coefficient, 49
a_0 bohr (unit of length), 3, 13, 16
a_0 Bohr radius, ifc, 6, **13**
a adsorbed (subscript), 48
a atom, **27**
a atto (SI prefix), 14
a symmetry label, 35
a year (unit of time), 16
aq aqueous solution, **48**, 49, 61
at atomization (subscript), **51**
ads, a adsorption (subscript), 48, **51**
app apparent (superscript), 51
atm atmosphere, 15, **17**

A magnetic vector potential, **24**
A absorbance, **38**, 39
A electron affinity, 27
A integrated absorption coefficient, 37
A activity (radioactive), 17, **28**
$A, \mathcal{A}$ affinity of reaction, **50**, 51
A area, 4, 16, **22**, 24, 27, 63, 64
A Einstein coefficient, Einstein transition probability, **38**, 40
A Helmholtz energy, **49**
A hyperfine coupling constant, 31
A nucleon number, mass number, **27**
A pre-exponential factor, 54, **55**
A rotational constant, 29
A spin-orbit coupling constant, 30
A_{r} relative atomic mass, **44**
A ampere (SI unit), 2, 11, 18
A symmetry label, 30, 34
Å ångström (unit of length), 16

b fundamental translation vector, **41**
b* reciprocal lattice vector, **41**
b molality, **45**
b Tafel slope, **60**
b unit cell length, lattice constant, lattice parameter, 41
b van der Waals coefficient, 49
b barn (unit of area), 16
b symmetry label, 35
bar bar (unit of pressure), **15**, 17

B magnetic flux density (magnetic induction), 19, 20, **24**, 31, 32
B Debye–Waller factor, 41
B Einstein coefficient, Einstein transition probability, **38**
B Napierian absorbance, **38**
B rotational constant, 29
B second virial coefficient, 49
Bi biot (unit of electric current), 18
Bq becquerel (SI unit), 17

c fundamental translation vector, **41**
c velocity, **22**, 43
c* reciprocal lattice vector, **41**
c amount concentration, 2, 5, 37–39, **44**, 48, 50, 52–57, 59, 60, 64
c specific heat capacity (constant pressure), **4**, 64, 65
c speed, 11, 13, **22**, 43, 65
c unit cell length, lattice constant, lattice parameter, 41
c speed of light in medium, 36
c_0 speed of light in vacuum, ifc, 6, **11**, 13, 20, 22, 27, 29, 36, 41
c_1 first radiation constant, ifc
c_2 second radiation constant, ifc
c centi (SI prefix), 14
c combustion (subscript), **51**, 52
c symmetry label, 35
cd candela (SI unit), 11, 36
cr crystalline, 3, **48**
cal calorie (unit of energy), 17
ccc critical coagulation concentration, 63
cmc critical miscellisation concentration, 63

C capacitance, 18, **24**
C heat capacity, 3, 4, **49**
C number concentration, **44**, 54, 55
C rotational constant, 29
C third virial coefficient, 49
C_n n-fold rotation symmetry operator, 34
C coulomb (SI unit), 11, 13, 18, 19, 24, 27, 28, 30, 31, 40
Ci curie (unit of radioactivity), 17
°C degree Celsius (SI unit), 3, 15, 17, 23, 38, 49

d centrifugal distortion constant, 30
d collision diameter, 55
d degeneracy, statistical weight, 29, 30, 37, **43**
d diameter, distance, thickness, 22, 23
d lattice plane spacing, 41
d relative density, 23
d day (unit of time), 16
d deci (SI prefix), 14
da deca (SI prefix), 14
dil dilution (subscript), 51
dpl displacement (subscript), 51
dyn dyne (unit of force), 17

$\boldsymbol{D}$ electric displacement, 19, 20, **24**
D absorbed dose of radiation, 17
D centrifugal dstortion constant, 30
D Debye–Waller factor, 41
D diffusion coefficient, 60, 64, 65
D dissociation energy, 13, 27
D_{AB} dipolar coupling constant, 31
D_{T} thermal diffusion coefficient, 64
D debye (unit of electric dipole moment), 18
Da dalton (unit of mass), 6, 12, 16, 44

$\boldsymbol{e}$ unit vector, 22
e elementary charge, ifc, 6, **11**, 27, 59
e linear strain, 23
e electron, 6
e energetic (subscript), 36
e symmetry label, 34
eV electronvolt (unit of energy), 6, 12, 15, 17
erg erg (unit of energy), 17, 19

$\boldsymbol{E}$ electric field strength, 5, 18, 20, **24**, 27, 28, 60
E activation energy, threshold energy, **55**
E cell potential, **59**, 61
E electric potential difference, **24**
E energy, 17, **23**, 26, 27, 29, 43
E identity symmetry operator, 33, 34
E irradiance, **36**
E modulus of elasticity, 23
E potential (electrochemistry), **59**, 60, 62
E^* space-fixed inversion symmetry operator, 33
E_{h} Hartree energy, 13, 15, 17, 27
E exa (SI prefix), 14
E excess quantity (superscript), 51
E symmetry label, 34
Ei exabinary (prefix for binary), 14

f activity coefficient, **50**
f atomic scattering factor, 41
f frequency, 22, 31
f friction factor, 23
f fugacity, **50**
f oscillator strength, 38, 40
$f(c_x)$ velocity distribution function, **43**
f femto (SI prefix), 14
f formation reaction (subscript), 3, 48, 50
f fluid phase, 48
ft foot (unit of length), 16
fus fusion, melting (subscript), 51

$\boldsymbol{F}$ Fock operator, **26**
$\boldsymbol{F}$ force, 17, 20, **23**
$\boldsymbol{F}$ angular momentum, **33**
F Faraday constant, ifc, **59**
F fluence, 37
F frequency, 31
F Helmholtz energy, 49
F rotational term, 29
F structure factor, 41
$F(c)$ speed distribution function, 43
F farad (SI unit), 18
F symmetry label, 34
°F degree Fahrenheit (unit of temperature), 17
Fr franklin (unit of electric charge), 18

g acceleration of gravity, **17**, 22, 23, 65
g degeneracy, statistical weight, 29, **43**
g, g_{e} g-factor, ifc, 27, **31**
g_{n} standard acceleration of gravity, ifc
g vibrational anharmonicity constant, 29
g gas, **48**, 51, 52, 61
g gram (unit of mass), 5, 14, 17, 67, 68
g gerade symmetry label, 6, 34

$\boldsymbol{G}$ reciprocal lattice vector, **41**
G (electric) conductance, **24**
G Gibbs energy, 3, 4, **49**–52, 55, 56, 59, 61
G gravitational constant, ifc, 23
G integrated (net) absorption cross section, 37–40
G shear modulus, 23
G thermal conductance, 64
G vibrational term, 29
G weight, 23
Gr Grashof number (mass transfer), 65

G gauss (unit of magnetic flux density), 19
G giga (SI prefix), 14
Gi gigabinary (prefix for binary), 14
Gy gray (SI unit), 17

h coefficient of heat transfer, **64**, 65
h height, 22
h index of lattice planes, 41, **42**
$h, \hbar$ Planck constant ($\hbar = h/2\pi$), ifc, 6, **11**, 13, 15–17, 19, 25–29, 31–33, 38, 40, 41, 43, 55–57
h hecto (SI prefix), 14
h hour (unit of time), **16**
hk metric horse power (unit of power), 17

$\boldsymbol{H}$ magnetic field strength, 19, 20, **24**
H Coulomb integral, 26
H dose equivalent, 17
H enthalpy, 3, 4, 48–52, 55, 56
H fluence, 37
H Hamilton function, Hamiltonian, **23**, **25**, 26
H Heaviside function, **9**, 43
Hz hertz (SI unit), 3, 11, 22, 29, 31, 32, 36

i inversion symmetry operator, 34
i square root of -1, **7**, 25, 34, 41
id ideal (superscript), 51
in inch (unit of length), 16
imm immersion (subscript), 51
iep isoelectric point, 63

$\boldsymbol{I}$ nuclear spin angular momentum, 31, **33**
I electric current, 2, 18, 20, **24**, 60
I intensity, 2, **36**, 37
I ionic strength, **50**, 59
I ionization energy, 27
I moment of inertia, **23**, 29

$\boldsymbol{j}$ internal vibrational angular momentum, 33
$\boldsymbol{j}$ electric current density, 20, **24**, 60

$\boldsymbol{J}$ angular momentum, 17, 33
$\boldsymbol{J}$ electric current density, **24**, 60
J angular momentum quantum number (component), 33–35, 56
J Coulomb integral, 26
J heat flux (density), **64**
J moment of inertia, 23
J_{AB} spin-spin coupling constant, 31
J joule (SI unit), 3, 4, 6, 11–13, 15, 17–19, 23–25, 27, 28, 31, 32, 36, 37, 41, 43, 49–52, 54–56, 59, 63
k absorption index, **38**
k angular momentum quantum number (component), 33
k, k_{B} Boltzmann constant, ifc, **11**, 43, 55, 56
k coefficient of heat transfer, 64
k coverage factor, 67, 68
k_{D} Debye angular wavenumber, **41**
k decay constant, 28
k index of lattice planes, 41, **42**
k rate coefficient, rate constant, 54–58, 60
k thermal conductivity, **64**
k_{d} mass transfer coefficient, **60**, 65
k_{H} Henry's law constant, **50**
k kilo (SI prefix), **14**
kg kilogram (Si unit), 5, 11–13, 16–19, 23, 27, 29, 30, 38, 43, 44, 45, 49, 50, 53, 59, 63, 64
kp kilopond (unit of force), 17
kgf kilogram-force (unit of force), 17

$\boldsymbol{K}$ rate (coefficient) matrix, **56**
K absorption coefficient, 38
K angular momentum quantum number (component), **33**, 35
K bulk modulus, **23**
K coefficient of heat transfer, 64
K equilibrium constant, 7, **50**, 52
K exchange integral, 26
K kinetic energy, **23**
K_{M} Michaelis constant, **56**
K_{AB} reduced nuclear spin-spin coupling constant, **31**
K kelvin (SI unit), 1, 3, **11**, 17, 37, 39, 41, 43, 49–52, 64
Ki kilobinary (prefix for binary), 14

$\boldsymbol{l}$ electron orbital angular momentum, **33**
$\boldsymbol{l}$ internal vibrational angular momentum, **33**
l cavity spacing, path length, 37–39
l index of lattice planes, 41, **42**
l length, 2, 7, 16, **22**, 65
l vibrational quantum number, 29, **33**
l liquid, 48
l litre (unit of volume), 4, 16

$\boldsymbol{L}$ angular momentum, 17, **23**, 30, 33
L Avogadro constant, ifc, 6, 44
L angular momentum quantum number, 33, 34
L_{D} Debye length, 63
L inductance, 19, **24**
L Lagrangian function, Lagrangian, **43**
L radiance, **36**
L langmuir (unit of pressure-time product), 64
L litre (unit of volume), 3, 16

$\boldsymbol{m}$ magnetic dipole moment, 13, 19, 20, 24, **27**
m angular momentum quantum number (component), **33**
m electric mobility, 60
m mass, 2, 4, 5, 12, 16, **23**, 25, 43, 44, 45, 49, 63, 64, 67, 68
m molality, **45**, 50, 52, 53, 59

m order of reaction, **54**
m_u atomic mass constant, ifc, **44**
m_e electron mass, ifc, 6, **13**, 15, 16, 19, 26, 27, 32, 38, 40
m_n neutron mass, ifc
m_p proton mass, ifc, 32
m metre (SI unit), 1, 3, 5, 11–13, 15, 16, 17–19, 22–25, 27–31, 36–38, 41, 43–45, 49, 50, 54–56, 59, 60, 63, 64
m milli (SI prefix), 14
mi mile (unit of length), 16
min minute (unit of time), 16
mix mixing (subscript), 51
mol mole (SI unit), 2–4, 11, 15, 19, 24, 37–39, 43–45, 48–56, 59, 60, 63
mmHg millimetre of mercury (unit of pressure), 17

$\boldsymbol{M}$ magnetization, 19, 20, **24**, 31
$\boldsymbol{M}$ torque, 23
$\boldsymbol{M}$ transition dipole moment, 30
M angular momentum quantum number (component), 33
M molar mass, **44**, 48
M_u molar mass constant, ifc, **44**
M_r relative molecular mass, 44
Ma Mach number, 65
M mega (SI prefix), 14
M molar (unit of concentration), **45**
Mi megabinary (prefix for binary), 14
Mx maxwell (unit of magnetic flux), 19

n amount of substance, chemical amount, 2, 4, 44, 45, 48, 49, 63
n charge number of cell reaction, 59, 60
n number of electrons, 26, 59, 60
n order of reaction, 56, 60
n principal quantum number, 27
n refractive index, **36**, 38
n nano (SI prefix), 14
n neutron, 6

$\boldsymbol{N}$ angular momentum, 33
N neutron number, **27**
N number of entities, 28, 41, **44**, 63
N number of events, 22, 38
N number of measurements, 67
N number of states, 43, **56**
N_A Avogadro constant, ifc, 6, **11**, 15, 39, 43, 44, 55, 57
Nu Nusselt number (mass transfer), 65
N newton (SI unit), **17**, 23, 24, 31, 63

Oe oerstedt (unit of magnetic field strength), 19

$\boldsymbol{p}$ electric dipole moment, 18, 20, **24**, 27, 28
$\boldsymbol{p}$ momentum, **23**, 25, 43
p fractional population, **37**, 43
p permutation symmetry operator, 33
p pressure, 1, 3–5, 15, 17, **23**, 37, 44, 48–55, 60, 63, 65
p probability, 43
p^* permutation-inversion symmetry operator, 33
p pico (SI prefix), 14
p photon (subscript), 36
p proton, 6
pH pH, 3, **59**

$\boldsymbol{P}$ density matrix, 26, **43**
$\boldsymbol{P}$ dielectric polarization, **20**, **24**
P heat flux, thermal power, **49**, **64**
P permutation symmetry operator, 33
P power, 17, **23**
P pressure, **23**, 44
P probability, **43**, 55
P probability density, **25**
P radiant power, **36**, 37
P transition probability, 55
P (volume) polarization, 19, 20
P weight, 23
P^* permutation-inversion symmetry operator, 33
Pe Péclet number (mass transfer), 65
Pr Prandtl number, 65
P peta (SI prefix), 14
P poise (unit of dynamic viscosity), 17
P symmetry label, 7, 34, 35, 47
Pa pascal (SI unit), 1, 15, 17, 23, 37, 39, 44, 49, 50, 53, 54, 62, 64
Pi petabinary (prefix for binary), 14

$\boldsymbol{q}$ electric field gradient, 28, **31**
$\boldsymbol{q}$ angular wave vector, 41
q_D Debye angular wavenumber, 41
q electric field gradient (element), **28**
q flux of mass, 64
q generalized coordinate, **22**, 23, 43
q heat, **49**
q partition function, 37, 56
q vibrationa normal coordinate, 30

$\boldsymbol{Q}$ quadrupole moment, 18, 27, 28, 31, 32
Q electric charge, 18, 20, **24**, 59, 60
Q heat, **49**, 64
Q partition function, **43**
Q radiant energy, 36
Q reaction quotient, **50**, 52
Q vibrational normal coordinate, 30
Q Q-branch label, 35

$\boldsymbol{r}$ position vector, 20, **22**, 24, 41, 43
r interatomic distance, **30**, 31
r internal vibrational coordinate, 30
r radius, **22**, 55
r rate of concentration change, **54**
r spherical coordinate, **22**, 25, 26
r reaction (subscript), 50, **51**, 59, 61
rad radian (SI unit), 16, 22, 32, 36, 38, 41
refl reflected (subscript), 36

$\boldsymbol{R}$ (ion) position vector, 41
$\boldsymbol{R}$ nuclear orbital angular momentum, 33
$\boldsymbol{R}$ transition dipole moment, 30
R electric resistance, 18, **24**
R molar gas constant, ifc, 5, **43**, 48–52, 55, 56, 59–61
R internal vibrational coordinate, 30
R position vector, **41**
R reflectance, 37
R resolving power, 37
R_∞ Rydberg constant, ifc, 27
R thermal resistance, 64
Re Reynolds number, 3, **65**
Ry rydberg (unit of energy), 17
°R degree Rankine (unit of thermodynamic temperature), 17

$\boldsymbol{s}$ electron spin angular momentum, 33
s length of path, length of arc, 22
s sedimentation coefficient, **63**
s solubility, **45**
s standard deviation, **67**
s symmetry number, 43
s^2 variance, **67**
s second (SI unit), 5, 11, 12, 13, 15, **16**, 17–19, 22–25, 27, 28, 31, 32, 36, 38, 40, 41, 43, 54–56, 60, 63, 64
s solid, 48, 61
s symmetry label, 30
sr steradian (SI unit), 22, 36
sln solution, 48
sol solution (subscript), 51
sub sublimation (subscript), 51

$\boldsymbol{S}$ Poynting vector, 20, **24**
$\boldsymbol{S}$ probability current density, 25
$\boldsymbol{S}$ scattering matrix, 55
$\boldsymbol{S}$ electron spin angular momentum, 30, 31, **33**, 34
S integrated absorption coefficient, **37**, 39
S area, **22**
S entropy, 3, 48–51, 55, 56
S overlap integral, **26**
S statistical entropy, **43**
S vibrational symmetry coordinate, 30
Sc Schmidt number, 65
Sh Sherwood number, 65
S_n rotation-reflection symmetry operator, 34
St Stanton number (mass transfer), 65

$\boldsymbol{t}$ Bravais lattice vector, 41
t Celsius temperature, 17, **49**
t time, 2, 11, 16, **22**–24, 28, 36–38, 49, 54, 56–58, 60, 64
$t_{1/2}$ half life, **55**
tr transmitted (subscript), 36
tp triple point (subscript), 51
trs transition (subscript), 51

$\boldsymbol{T}$ torque, **23**
T kinetic energy, 12, **23**, 25
T period, characteristic time interval, 22
T relaxation time, **22**, 32
T thermodynamic temperature, 1–5, 17, 37, 39, 43, 48–52, 54–56, 59–61, 63–65
T total term, electronic term, 15, 29
T transmittance, 37
T tera (SI prefix), 13
T tesla (SI unit), 19, 24, 27, 31, 32
Ti terabinary (prefix for binary), 14
Torr torr (unit of pressure), 15, 17, 64

$\boldsymbol{u}$ displacement vector, 41
$\boldsymbol{u}$ velocity, **22**, 43
u electric mobility, 60
u standard uncertainty, 7, 67, 68
u lattice direction index, 42
u speed, **22**
u ungerade symmetry label, 34
u unified atomic mass unit, 6, 12, 16, **44**

U cell potential, 59
U electric potential difference, **24**
U electrode potential, 59
U expanded uncertainty, 67, 68
U internal energy, 49, 55

$\boldsymbol{v}$ velocity, 20, **22**–24, 43, 60, 64
v lattice direction index, 42
v rate of reaction, **54**, 58
v specific volume, volume, **4**, 22, 23
v speed, 12, 17, **22**, 63–65
v vibrational quantum number, **29**, 33, 35, 47
v visible (subscript), 36
vap vaporization (subscript), 51

V electric potential, 18, **24**
V limiting rate, **56**
V potential energy, **23**, 25, 27, 31
V volume, 4, 5, 16, 22–24, 27, 28, 36, 44, 48, 49, 54–56

V_{m} molar volume (ideal gas, 100 kPa, 0 °C), ifc
V volt (SI unit), 6, 12, 13, 18, 19, 24, 28, 59, 60

$\boldsymbol{w}$ velocity, **22**
w lattice direction index, 42
w mass fraction, **44**
w radiant energy density, 36
w speed, **22**
w work, **23**, **49**

W degeneracy, statistical weight, **43**
W number of states, **43**, **56**
W radiant energy, **36**
W weight, 23
W work, **23**, **49**
We Weber number, 65
W watt (SI unit), 17, 23, 24, 36, 37, 49, 64
Wb weber (SI unit), 19, 24

x amount fraction, **44**, 50, 53
x Cartesian coordinate, **22**, 23, 25
x energy parameter, 26
x fractional coordinate, 41
$\bar{x}$ mean, 67
x specific quantity, 49
x vibrational anharmonicity constant, 29

$\bar{X}$ molar quantity, 49
X reactance, **24**

y amount fraction (gas), **44**, 52
y Cartesian coordinate, **22**, 23, 25
y fractional coordinate, 41
y yocto (SI prefix), 14
yd yard (unit of length), 16

Y spherical harmonic function, **25**
Y yotta (SI prefix), 14
Yi yottabinary (prefix for binary), 14

z Cartesian coordinate, **22**, 23, 25
z charge number, 50, 59–62
z collision frequency, collision frequency factor, 55
z cylindrical coordinate, 22
z fractional coordinate, 41
z partition function, **43**
z zepto (SI prefix), 14

Z collision density, collision number, 55
Z compression factor, **49**
Z partition function, **43**
Z proton number, atomic number, 26, **27**
Z zetta (SI prefix), 14
Zi zettabinary (binary prefix), 14

$\boldsymbol{\alpha}$ electric polarizability, 18, **28**
α absorptance, **37**, 39
α absorption coefficient, 38
α angle of optical rotation, **38**
α coefficient of heat transfer, 64
α Coulomb integral, 26
α cubic expansion coefficient, 65
α degree of reaction, 45
α fine-structure constant, ifc, 13
α optical rotatory power, 38
α plane angle, 16, **22**
α polarizability, 18, **28**
α spin wavefunction, **25**, 26
α unit cell angle, 41
α α-particle, 6
α phase, 48, 59

$\boldsymbol{\beta}$ hyper-polarizability, 28
β degeneracy, statistical weight, **43**
β plane angle, **22**
β reciprocal energy parameter (to replace temperature), 43
β resonance integral, 26
β spin wavefunction, **25**, 26
β unit cell angle, 41
β phase, 48, 59

$\boldsymbol{\gamma}$ hyper-polarizability, 28
γ activity coefficient, **50**, 52, 59
γ gyromagnetic ratio, **27**, 31, 32
γ mass concentration, 38, **44**
γ plane angle, 16, **22**
γ ratio of heat capacities, 49
γ shear strain, 23
γ surface tension, **23**, 63, 65
γ transmission coefficient, **56**
γ unit cell angle, 41

$\mathit{\Gamma}$ integrated absorption coefficient, **37**, 39
$\mathit{\Gamma}$ level width, linewidth, 56, 57
Γ gamma function, 9

δ centrifugal distortion constant, 30
δ chemical shift, 32
δ thickness, **22**, 60
δ Dirac delta function, Kronecker delta, 7, **8**, 26, 41

$\mathit{\Delta}$ centrifugal distortion constant, 30
Δ change in extensive quantity, **50**–52
Δ finite change, **8**, 23, 35

ε_{F} Fermi energy, 41
ε linear strain, 23

ε molar decadic absorption coefficient, 4, 38, 39
ε orbital energy, 26
ε permittivity, **24**
ε_0 permittivity of vacuum, electric constant, ifc, 13, 18, 20, **24**, 38
$\mathrm{\varepsilon}$ Levi-Civita symbol, **9**
$\mathrm{\varepsilon}$ unit step function, Heaviside function, **9**

ζ bulk viscosity, **64**
ζ Coriolis ζ-constant, **29**, 33
ζ electrokinetic potential, 60

η overpotential, 60
η viscosity, 17, **23**, 64, 65

ϑ cylindrical coordinate, 22
ϑ plane angle, 22
ϑ volume strain, bulk strain, 23

θ Bragg angle, 41
θ Celsius temperature, **49**
θ contact angle, 63
θ spherical polar coordinate, **25**
θ surface coverage, 63
θ internal vibrational coordinate, 30

$\boldsymbol{\Theta}$ quadrupole moment, **18**
Θ_{D} Debye temperature, **41**
Θ plane angle, 36

κ asymmetry parameter, 30
κ conductivity, 4, 18, **20**
κ magnetic susceptibility, **19**
κ molar Napierian absorption coefficient, 37, **38**
κ ratio of heat capacities, 49
κ transmission coefficient, **56**

λ absolute activity, **49**, 50
λ angular momentum quantum number (component), 33
λ decay constant, 28
λ molar conductivity of an ion, 48, **60**
λ thermal conductivity, **64**
λ wavelength, 1, 36, 38, 41

Λ angular momentum quantum number (component), 33, 34
Λ molar conductivity, 4, 48, **60**

$\boldsymbol{\mu}$ electric dipole moment, 18, **27**, 38, 40
μ chemical potential, 3, 41, **49**, 50–53
$\tilde{\mu}$ electrochemical potential, 59
μ dynamic friction factor, 23
μ Joule–Thomson coefficient, **49**
μ magnetic dipole moment, 19, **24**, 27, 31
μ permeability, 6, **24**
μ reduced mass, 6, **23**, 55, 56
μ viscosity, **23**
$\tilde{\mu}$ electrochemical potential, 59
μ_{e} electron magnetic moment, ifc
μ_0 permeability of vacuum, magnetic constant, ifc, 20, **24**, 31
μ_{B} Bohr magneton, ifc, 13, 19, **27**, 31, 32
μ_{N} nuclear magneton, 19, **27**
$\mathrm{\mu}$ micro (SI prefix), 14
$\mathrm{\mu}$ micron (unit of length), 16
$\mathrm{\mu}$ muon, 6

ν charge number of cell reaction, 59
ν_{D} Debye frequency, 41
ν frequency, **22**, 27, 29, 31, 32, 36–38, 56, 57
ν kinematic viscosity, **23**
ν stoichiometric number, **45**, 50–52, 54, 59–61
$\tilde{\nu}_{\mathrm{D}}$ Debye wavenumber, 41
$\tilde{\nu}$ wavenumber (in vacuum), 15, **29**, 36–41
v vibrational state symbol, 35

ξ extent of reaction, advancement, **45**, 50, 51, 54, 57
ξ magnetizability, 19, **27**

Ξ standardized resonance frequency, 32

$\boldsymbol{\pi}$ internal vibrational angular momentum, 33
π surface pressure, 63
$\mathrm{\pi}$ symmetry label, 35
$\mathrm{\pi}$ ratio of circumference to diameter, 7, 13, 16–20, 22, 24, 27, 29, 31, 34, 36, 38, 40, 41, 43, 55

Π product sign, 8, 50, 51, 54, 60
Π symmetry label, 7, 35

$\boldsymbol{\rho}$ density matrix, 43
ρ charge density, 18, 20, **24**, 27
ρ cylindrical coordinate, 22
ρ density of states, **43**, 56
ρ density operator, 43
ρ radiant energy density, **36**
ρ mass density, mass concentration, 4, **23**, 38, 44, 64, 65
ρ reflectance, 37
ρ_A surface density, 23

σ absorption cross section, 37, 38
σ area per molecule, 63
σ cross section, 37, 38, 55
σ conductivity, electrical conductivity, 18

σ density operator, 43
σ normal stress, 23
σ reflection symmetry operator, 34
σ shielding constant, 32
σ spin quantum number (component), 33
σ Stefan–Boltzmann constant, 37
σ surface tension, **23**, 63
σ symmetry number, 43
$\mathrm{\sigma}$ symmetry label, 35

Σ spin quantum number (component), 33, 34
Σ summation sign, 8, 23, 26, 29, 37, 38, 40, 41, 43–45, 51, 54, 59–61, 67, 68

τ characteristic time interval, relaxation time, 22, **55**
τ correlation time, 32
τ mean life, lifetime, 55, 56
τ shear stress, **23**, 64
τ transmittance, 37

φ fluidity, 23
φ plane angle, 22
φ spherical coordinate, 22

ϕ electric potential, 18, 20, **24**
ϕ fugacity coefficient, 50
ϕ inner electric potential, 59
ϕ molecular orbital, 26
ϕ quantum yield, 56, 57
ϕ spherical polar coordinate, 25
ϕ volume fraction, **44**
ϕ wavefunction, 25, 43

Φ configuration function, **26**
Φ heat flux, thermal power, 49, 64
Φ magnetic flux, 19, **24**
Φ potential energy, 23
Φ quantum yield, 56
Φ radiant power, 36
Φ work function, **41**

$\boldsymbol{\chi}$ quadrupole interaction energy tensor, **28**, 32
χ atomic orbital, **26**
χ electronegativity, 27
χ magnetic susceptibility, 19, 20, **24**
χ_e electric susceptibility, 20, **24**
χ_m molar magnetic susceptibility, 19, **24**

ψ wavefunction, 25, 30

Ψ wavefunction, 25, 26, 43

ω angular frequency (velocity), **22**, 27, 36
ω_D Debye angular frequency, 41
ω degeneracy, statistical weight, 43
ω harmonic (vibrational) wavenumber, 29
ω solid angle, 22

Ω angular momentum quantum number (component), 33, 34
Ω nutation angular frequency, 31
Ω partition function, 43
Ω solid angle, 22
Ω volume in phase space, 43
$\mathrm{\Omega}$ ohm, 18

Special symbols

$^\circ$	degree (unit of arc), <u>16</u>, 38
$^{\ominus}$, $^\circ$	standard (superscript), 49, 51–53, 55, 56, 59, 60
*	excitation, 46, 57
*	pure substance (superscript), 51, 52
‡, ≠	activated complex (superscript), transition state (superscript), 51, 55, 56
∞	infinite dilution (superscript), 51
$^\mathrm{e}$	even parity symmetry label, 34
$^\mathrm{o}$	odd parity symmetry label, 34
[B]	amount concentration, concentration of B, **44**, 48, 58
Δ_r	derivative with respect to extent of reaction, **50**, 51, 61
$\dim(Q)$	dimension of quantity Q, 2
∇	nabla, 9, 20, 24–26, 64
$[\alpha]_\lambda^\theta$	specific optical rotatory power, 38
$[Q]$	unit of quantity Q, 1
$\{Q\}$	numerical value of quantity Q, 1

9 SUBJECT INDEX

When more than one page reference is given, bold print is used to indicate the general (defining) reference. Underlining is used to indicate a numerical entry of the corresponding physical quantity. Greek letters are ordered according to their spelling and accents are ignored in alphabetical ordering. Plural form is listed as singular form, where applicable. The acronym 'ifc' for a page number refers to the inside front cover.

absolute activity, **49**
absorbance, 39, **39**
    decadic, **38**
    Napierian, **38**
absorbed dose (of radiation), 17
absorbing path length, 39
absorptance, **37**, 39
    internal, 39
absorption, 35, 38, **39**
    rate of photon, 57
absorption band, 37
absorption band intensity
    net integrated, 39
absorption coefficient, 37, 38
    decadic, 38
    linear decadic, 38
    linear Napierian, 38
    molar, 4
    molar decadic, **38**, 39
    molar Napierian, 38
    Napierian, 38, **38**
    net, 37
absorption cross section
    integrated, 38
    integrated net, 38
    net, 37
absorption factor, 37
absorption index, 38
absorption intensity, 39
absorption line, 37
    Lorentzian, 57
absorption spectrum, 37
acceleration, 17, **22**
acceleration of gravity, 22, **63**, 65
    standard, 17
acid dissociation
    equilibrium constant for an, 50
action, 13
    au of, <u>17</u>
activated complex (superscript), 51
activation
    standard enthalpy of, 55
    volume of, 55
activation energy, 55
    Arrhenius, 55
activity, 49, **49**, 52, 59, 60
    absolute, **49**
    catalytic, 58
    mean ionic, 59
    radioactive, 17, **28**
    relative, **49**
    unit, 62
activity coefficient, **50**, 59
    mean ionic, 59
activity coefficient (concentration basis), 50
activity coefficient (Henry's law), 50
activity coefficient (molality basis), 50
activity coefficient (amount fraction basis), 50
activity coefficient (Raoult's law), 50
activity of an electrolyte, 59
adjoint, *see* Hermitian conjugate
adsorbed amount, 63
adsorbed molecules
    number of, 63
adsorbed species, 48
adsorption, 63
adsorption (subscript), 51
affinity
    electron, 27
affinity of reaction, **50**, 51
α-particle, 6
amount, 2, 44, **47**, 48
    chemical, 2, **47**
    excess, 63
    surface excess, **63**
amount concentration, 2, 37–39, **44**, 48, 52, 54
amount fraction, **44**, 50, 52, 53
amount of substance, 2, 11, 47, **47**, 48, 49
amount of substance of photons, 57
amount-of-substance concentration, 2, 38, 39
amount-of-substance fraction, **44**
Ampère law, 20
ampere, 2, 11, 18
angle, 36
    Bragg, 41
    contact, 63
    plane, **22**, 38
    reciprocal unit cell, 41
    solid, **22**, 36
    unit cell, 41
angle of optical rotation, 38, **38**
ångström, <u>16</u>

angular frequency, **22**, 36
  Debye, 41
  Larmor, 27
  nutation, 31
angular frequency unit, 31
angular fundamental translation vector, 41
angular momentum, 13, 17, **23**, 26, 31, 33
  electron orbital, 33
  electron orbital plus spin, 33
  electron spin, 33
  internal vibrational, 33
  nuclear orbital, 33
  nuclear spin, 33
  orbital, 34
  spin, **25**, 31
  total, 34
angular momentum component, 33
angular momentum eigenvalue equation, 33
angular momentum operator, 25, **33**
  symbol of, 33
angular momentum quantum number, **33**
angular reciprocal lattice vector, 41
angular velocity, **22**
anharmonicity constant
  vibrational, 29
anode, 61
anodic partial current, 60
anticommutator, 25
apparent (superscript), 51
aqueous solution, 48, **48**, 49, 62
aqueous solution at infinite dilution, **48**
arc
  length of, 22
area, 4, 16, **22**, 27, 64
  infinitesimal, 22
  specific surface, 63
  surface, **63**
  vector element of, 24
area element, 36
area per molecule, 63
areic, 4
areic charge, 4
Arrhenius activation energy, 55
astronomy, 12
asymmetric reduction, 30
asymmetric top, **33**, 35
asymmetry parameter, 30
atmosphere, 15
  standard, 17
atmospheric pressure, 49
atomic basis function, 26
atomic mass, **27**, 44
  relative, **44**
atomic mass constant, ifc, 44
atomic mass unit
  unified, ifc, 6, 12, 16, 44
atomic number, 27, **27**, 46
atomic scattering factor, 41
atomic spectroscopy, 23
atomic state, 34
atomic unit, 2, 13, 15–17, 26
  symbol for, 13
atomic unit of magnetic dipole moment, 13
atomic weight, **44**
atomization, 52
atomization (subscript), 51
atto, 14
au of action, 13, 17
au of force, 13, 17
au of speed, 13, 17
au of time, 13, 16
average, **45**
average collision cross section, 55
average speed, **43**
averaged collision cross section, 55
Avogadro constant, ifc, 2, 6, 11, 15, 43, 44, 48, 55
Avogadro number, 44
axis of quantization, 33
azimuthal quantum number, 33

band intensity, 37, 39
bar, **15**, 17
barn, 16
base hydrolysis, 50
base quantity, **2**, 11
  symbol of, 2
base unit, 11, 12, 36
  SI, 2, 11, 14
basis function
  atomic, 26
becquerel, 17
Beer–Lambert law, **36**, 37
bimolecular reaction, **57**, 58
binary multiple, 14
binomial coefficient, 8
biochemical standard state, 53
biot, 18
BIPM, **3**, 6
black body, 37
bohr, 13, 16
Bohr magneton, ifc, 13, 19, **27**, 32
Bohr radius, ifc, 6
Boltzmann constant, ifc, 11, **43**
bond length
  equilibrium, 13
Bragg angle, 41
Bragg reflection, 42
branches (rotational band), 35
Bravais lattice vector, 41
bulk modulus, 23
bulk phase, 59, 63
bulk strain, 23
bulk viscosity, 64

- calorie
  - 15 °C, 17
  - international, 17
  - thermochemical, 17
- candela, 11, 36
- canonical ensemble, **43**
- capacitance, 18, **24**
- Cartesian component, 9
- Cartesian coordinate system, 9
- Cartesian (space) coordinate, **22**
- catalytic activity, 58
- catalytic reaction mechanism, 57
- cathode, 61
- cathodic current, 59
- cathodic partial current, 60
- cavity, 37
  - Fabry–Perot, 37
  - laser, 37
- cell
  - electrochemical, 59, 61
  - electrolytic, 61, **61**
- cell diagram, 61, **61**
- cell potential, **59**
- cell reaction
  - electrochemical, **61**
- Celsius temperature, 17, **49**
- centi, 14
- centigrade temperature, 49
- central Coulomb field, 27
- centre of mass, 34
- centrifugal distortion constant, 30
- centrifugation, 63
- CGPM, viii
- cgs unit, 13, 16–17
- chain-reaction mechanism, 57
- characteristic Debye length, 63
- characteristic number, 3
  - transport, 65
- charge, 13, **24**, 26, 60
  - areic, 4
  - electric, 18, **24**
  - elementary, 6, 12, 13, 27
  - point of zero, 63
- charge (au)
  - proton, 18
- charge density, 18, **24**, 27
  - surface, 4
- charge number, 26, 46, 59
  - ionic, 46
- charge number of cell reaction, 59
- charge transfer, 60, 61
- chemical amount, 2
- chemical composition, 44, **47**
- chemical-compound name, 6
- chemical element, 45
  - symbol for, 6, **45**, 46
- chemical energy, 61
- chemical equation, 45
  - bimolecular elementary reaction, 58
  - elementary reaction, **57**, 58
  - general, 45
  - stoichiometric, **47**, **51**, 54, 58
  - unimolecular elementary reaction, 58
- chemical equilibrium, **61**
- chemical formula, 45–47
- chemical kinetics, **54**, 57
- chemical potential, 3, 4, **49**, 52
  - standard, **49**, 52, 53
- chemical reaction, 51, 54, 61
  - equation for, 47
  - transition state theory, 55
- chemical shift, 32
- chemical substance, 47
- chemical symbol for element, 46
- chemical thermodynamics, 49–53
- chemistry
  - quantum, 25
- chemistry (quantum), 13
- choice of standard state, 50
- choice of unit, 1
- CIPM, 6
- circular frequency
  - Larmor, 27
- classical mechanics, 23
- closed-shell systems, 26
- CODATA, **13**
- coefficient
  - absorption, 37, 38
  - activity, **50**, 59
  - Einstein, **38**
  - extinction, 39
  - Joule–Thomson, 49
  - mass transfer, **60**
  - molar absorption, 4
  - pressure virial, 49
  - sedimentation, 63, **63**
  - transmission, 56
  - van der Waals, 49
  - virial, 49
- coefficient of heat transfer, 64, 65
- coherent system, 12
- coherent system of units, **12**
- coherent unit, 12
- collision, 55
- collision cross section, 55
  - averaged, 55
- collision density, 55
- collision diameter, 55
- collision energy, 56
- collision frequency, **55**
- collision frequency factor, 55
- collision model

  hard sphere, 55
collision number, 55
collision partner, 54
collisions
  total number of, 55
colloid chemistry, 63, **63**
combined standard uncertainty, 67, 68
combustion, 52
  standard enthalpy of, 52
combustion reaction (subscript), 51
commutator, 25
complex conjugate, 25
complex conjugate matrix, 9
complex mechanism, 57
complex refractive index, **38**
component quantum number, 33
composite mechanism, 57
compressibility factor, 49
compression factor, 49
compression modulus, 23
concentration, 50, 52–54, 59, 60
  amount, 2, 37, 38, **44**, 48, 52, 54
  amount-of-substance, 2, 38, 39
  critical coagulation, 63
  critical micellisation, 63
  excited state, 57
  mass, 38, **44**
  mean ionic, 59
  number, 44, 54, 55
  standard, **53**
  substance, 2
  surface, 54, **63**
  surface excess, 63
concentration at the interface, 60
concentration change
  rate of, 54
concentration perturbation, 55
concentration vector, 56
condensed phase, 49
conductance, 24
  thermal, 64
conductivity, 18, 20, **24**, 60
  ionic, 48, **60**
  molar, 4, 48, **60**
  thermal, **64**
conductivity of an ion
  molar, 60
conductor
  electron, 61
  ionic, 61
  metallic, 61
confidence
  level of, **68**
configuration function, 26
configuration state, 26
configuration state energy, **26**
conjugate transpose, 9
constant
  atomic mass, ifc, 44
  Avogadro, ifc, 2, 6, 11, 15, 43, 44, 48, 55
  Boltzmann, ifc, 11, **43**
  Coriolis coupling, 29
  Coriolis ζ-, **29**, 33
  effective Hamiltonian, 29
  electric, ifc, **24**
  Faraday, ifc, **59**
  fine-structure, ifc, 13
  fundamental physical, 6, 13
  gravitational, ifc
  Henry's law, 50, **50**
  hyperfine coupling, 31
  Josephson, ifc
  magnetic, ifc, **24**
  mathematical, 5, 7
  Michaelis, 56, **56**
  molar mass, ifc, **44**
  Planck, ifc, 6, 11, 13
  rate, 54, 55, 58
  rotational, 29
  shielding, 32
  Stefan–Boltzmann, ifc, **37**
  von Klitzing, ifc
contact angle, 63
conventions in chemical thermodynamics, 51
conventions in electrochemistry, 61
conversion
  rate of, **54**, 57
conversion factors, **16**
conversion of units, 15
conversion tables for units, 16–19
convolution of function, 9
coordinate, 41
  Cartesian (space), **22**
  cylindrical, 22
  dimensionless vibrational, 30
  electron, **26**
  fractional, 41
  generalized, **22**, 23, 43
  internal vibrational, 30
  mass adjusted vibrational, 30
  molecule-fixed, 34
  space-fixed, 33
  spherical polar, **22**
  vibrational normal, 30
coordinate representation, 25
core Hamiltonian, 26
Coriolis coupling constant, 29
Coriolis ζ-constant, **29**, 33
correlation time, 32
coulomb, 18
Coulomb field
  central, 27

Coulomb integral, 26
Coulomb's modulus, 23
coupling constant
Coriolis, 29
dipolar, 31
hyperfine, 31
nuclear quadrupole, 32
nuclear spin-spin, 31
quadrupole, 31, 32
reduced, 31
reduced nuclear spin-spin, 31
spin-orbit, 30
spin-rotation, 31
coupling path, 31
coverage factor, **67**, 68
critical coagulation concentration, 63
critical micellisation concentration, 63
cross product, 9
cross section
averaged collision, 55
collision, 55
integrated absorption, 38
integrated net absorption, 38
net absorption, 37
crystal face, 42
crystal lattice, 41
crystalline, **48**
cubic expansion coefficient, 65
cubic metre, 16
cumulative number of states, 43
curie, 17
curl of a vector field, 9
current
anodic partial, 60
cathodic, 59
cathodic partial, 60
electric, 2, 11, 18, **24**, 60
exchange, 60
Faradaic, 60
current density, 20, **60**
electric, **24**, 60
exchange, 60
probability, 25
cylindrical coordinate, 22

dalton, 6, 12, 16, 44, 45
day, 16
debye, 18
Debye angular frequency, 41
Debye angular wavenumber, 41
Debye frequency, 41
Debye length, 63
characteristic, 63
Debye screening length, 63
Debye temperature, **41**
Debye wavenumber, 41
Debye–Waller factor, 41

deca, 14
decadic absorbance, **38**
decadic absorption coefficient, 38
decadic logarithm, 54
decay
exponential, 57
decay constant, 28
decay rate constant, 28
decay time, 55
deci, 14
decimal marker, 7
decimal multiple of units, 4, **14**
decimal sign, 7
decimal submultiple of units, 4, **14**
degeneracy, 29, 37, **43**
degenerate modes, 29
degree, 16
degree Celsius, 17
degree of freedom, 67
degree of plane angle, 38
degree of reaction, 45
degree Rankine, 17
del operator, 9
density, **23**, 64, 65
charge, 18, **24**, 27
collision, 55
current, 20, **60**
electric current, **24**, 60
energy, **38**
flux, 64
ion, 41
mass, 4, **23**, 38, 47
probability, **25**
probability current, 25
relative, 23
surface, 23
density matrix, 43
density matrix element, 26, 43, **43**
density of states, **43**, 56
density operator, 43
deposition, 63
derived quantity, 2
determinant of a square matrix, 9
diameter, **22**
collision, 55
dielectric polarization, 20, **24**
diffusion coefficient, 60, 64, 65
thermal, 64
diffusion layer thickness, 60
diffusivity
thermal, 64, 65
dilution
infinite, **48**, 52
dilution (subscript), 51
dimension, 2
symbol of, 2

dimension one, 65
dimensionless, 22
dimensionless quantity, 3, 21
dimensionless vibrational coordinate, 30
dipolar coupling constant, 31
dipole, 20, **24**
    magnetic, 20
dipole moment, **24**, 28, 31
    electric, 13, 18, 24, 27
    magnetic, 13, 19, 24, **27**
dipole moment (molecule)
    transition, 30
dipole vector, 24
Dirac delta distribution, 9
displacement
    electric, 19, 20, **24**
    vibrational, 30
displacement (subscript), 51
displacement vector, 41
dissociation energy, 13, **27**
dissociation energy (ground state), 27
dissociation energy (potential minimum), 27
dissolution, 50
distance, **22**, 23, 24, 26, 41
    equilibrium, 30
    ground-state, 30
    interatomic, 30
    interatomic equilibrium, 30
    interatomic (internuclear), 30
distribution function
    speed, 43
divergence of a vector field, 9
dose equivalent, 17
dose of radiation
    absorbed, 17
duration, 22
dynamic friction factor, 23
dynamic viscosity, 17, **23**
dyne, 17

effective Hamiltonian constant, 29
effective magnetic flux density, 32
effusion
    rate of, 64
eigenvalue equation
    angular momentum, 33
einstein, 57
Einstein coefficient, **38**
Einstein transition probability, 39
elasticity
    modulus of, 23
electric charge, 18, **24**
electric constant, ifc, **24**
electric current, 2, 11, 18, **24**, 60
electric current density, **24**, 60
electric dipole moment, 13, 18, 24, 27
electric displacement, 19, 20, **24**
electric field gradient, **31**
electric field gradient tensor, 28
electric field strength, 5, 13, 18, 20, **24**, 28, 60
electric mobility, 60
electric polarizability, 13
electric polarizability of a molecule, 28
electric potential, 18, **24**
electric potential difference, **24**, 61
electric quadrupole moment, **13**, 18
electric resistance, 18, **24**
electric susceptibility, 20, **24**
electric tension, **24**
electrical energy, 61
electricity, 24
electrochemical cell, 59, 61
electrochemical cell reaction, **61**
electrochemical potential, 59
electrochemical reaction, 59–61, **62**
electrochemistry, 59–60
    conventions in, 61
electrode, 61
electrode potential, **59**
    equilibrium, **60**
    standard, **59**, 62
electrode reaction, 59, 60, **61**
electrokinetic potential, 60
electrolytic cell, 61, **61**
electromagnetic potential, **20**
electromagnetic radiation, 36
    intensity of, 36
electromagnetic theory, 20
electron, 6
electron affinity, 27
electron conductor, 61
electron configuration, 6, **34**
electron coordinate, **26**
electron energy, 41
electron magnetic moment, ifc
electron mass, ifc, 6, 12, **13**, 16, 32
electron number, 59
electron number (of an electrochemical reaction), 59
electron orbital, 31
electron orbital angular momentum, 33
electron orbital operator, 30
electron orbital plus spin angular momentum, 33
electron paramagnetic resonance, 31
electron spin angular momentum, 33
electron spin multiplicity, 34
electron spin operator, 30, **31**
electron spin resonance, 31
electron spin wavefunction, 25
electron work function, 41
electronegativity, 27
electronic term symbol, 47
electronic transition, 38
electronvolt, 6, 15, 17

element
  symbol for chemical, **45**, 46
elementary charge, ifc, 6, 12, 13, 27
elementary entity, 2, **47**, 48
elementary particles
  symbols for, 6
elementary reaction, 47, 55, **57**, 58
elementary step, 47
emission, **35**
  induced, 37, **38**
  spontaneous, **38**, 57
  stimulated, 38
emu, **16**, 18, 19
energy, 13, 17, **23**, 26, 27, 35, 43, 49
  activation, 55
  Arrhenius activation, 55
  chemical, 61
  collision, 56
  configuration state, **26**
  dissociation, 13, **27**
  electrical, 61
  electron, 41
  excitation, 15, **15**
  Fermi, 41
  Gibbs, 4, **49**, 61
  Hartree, ifc
  Helmholtz, **49**
  internal, **49**
  ionization, 27
  kinetic, 12, **23**
  orbital, 26
  partial molar Gibbs, 4, 59
  potential, **23**
  radiant, 36
  standard Gibbs, 59
  standard reaction Gibbs, 50
  threshold, 56
energy density, **38**
  radiant, 36
  spectral radiant, 36, **36**
energy flow
  rate of radiation, 20
energy level, 37
energy of activation
  standard Gibbs, 56
energy parameter, 26
  reciprocal, 43
energy per area, 37
energy tensor
  nuclear quadrupole interaction, 28
enplethy, **2**
ensemble, **43**
  canonical, **43**
  microcanonical, 43
enthalpy, 2, 4, **49**
  molar, **4**
  standard partial molar, 50
  standard reaction, 3, **50**
enthalpy of activation
  standard, 55
enthalpy of combustion
  standard, 52
enthalpy of formation
  standard, 48
  standard molar, 50
enthalpy of reaction
  molar, 51
enthalpy of vaporization
  molar, 51
entitic, 4
entity, 45–48, 60
  elementary, 2, **47**, 48
  formula, 48
  mass of, 44
entropy, 2, **49**
  equilibrium (maximum), 43
  molar, **48**
  standard reaction, 50
  statistical, 43
entropy change
  rate of, 54
entropy of reaction
  molar, 51
enzyme, 56
enzyme catalysis, 56
enzyme-substrate complex, 56
EPR, 31
equation
  stoichiometric, 45, **47**, 51, 54, 58
equations of electromagnetic theory, 20
equilibrium, 47, 50, 61
  thermal, 37
equilibrium (maximum) entropy, 43
equilibrium bond length, 13
equilibrium constant, 50
  standard, 52
  thermodynamic, 50
equilibrium constant (acid dissociation), 50
equilibrium constant (base hydrolysis), 50
equilibrium constant (concentration basis), 50
equilibrium constant (dissolution), 50
equilibrium constant (fugacity basis), 50
equilibrium constant (molality basis), 50
equilibrium constant (pressure basis), 50
equilibrium constant (water dissociation), 50
equilibrium distance, 30
  interatomic, 30
equilibrium electrode potential, **60**
equilibrium macroscopic magnetization, 31
equilibrium position vector, 41
equilibrium potential, **59**, 60
erg, 17, 17, 19

ESR, 31
estimated probability, 68
esu, **16**, 18, 19
evaporation (subscript), 51
exa, 14
exabinary, 14
exbi, 14
excess amount, 63
excess quantity (superscript), 51
exchange current, 60
exchange current density, 60
exchange integral, 26
excitation, 46
excitation energy, 15, **15**
  molar, 15
excitation symbol, 46
excited electronic state, 6, **46**
excited state, 35
excited state concentration, 57
excited-state label, 35
expanded uncertainty, 67, 68
expansion coefficient
  cubic, 65
expectation value (operator), 25
exponential decay, 57
extensive, 4
extensive quantity, **4**, 49, 64
extensive quantity (divided by area), 64
extensive quantity (divided by volume), 64
extensive thermodynamic quantity, 51
extent of reaction, 45, **45**, 51
extinction, 39
extinction coefficient, 39

Fabry–Perot cavity, 37
factor
  absorption, 37
  atomic scattering, 41
  collision frequency, 55
  compressibility, 49
  compression, 49
  Debye–Waller, 41
  dynamic friction, 23
  frequency, 55
  normalization, **25**
  pre-exponential, 54, **55**
  reflection, 37
  structure, 41
  transmission, 37
factorial, 8
Fahrenheit temperature, <u>17</u>
farad, <u>18</u>
Faradaic current, 60
Faraday constant, <u>ifc</u>, **59**
Faraday induction law, 20
femto, 14
Fermi energy, 41
field
  electric, 13
  magnetic, 19, 32
field gradient
  electric, **31**
field gradient tensor, 32
field strength
  electric, 5, 18, 20, **24**, 28, 60
  magnetic, 19, 20, **24**
fine-structure constant, <u>ifc</u>, 13
first-order kinetics
  generalized, 56
first-order rate constant, 56, **56**
first-order reaction, 60
first radiation constant, <u>ifc</u>
flow, 64
flow rate, 64
fluence, 37
fluence rate, 37
fluid phase, 48
fluidity, 23
fluorescence, 57
fluorescence rate constant, **56**
flux, 64
  heat, 64
  magnetic, 19, **24**
flux density, 60, 64
  (magnetic), 31
  effective magnetic, 32
  heat, **64**
  magnetic, 13, 19, 20, **24**, 31
  number, 64
flux density of mass, 64
flux of mass, 64
Fock operator, 26
  matrix element of, 26
fonts for symbol, 5
foot, <u>16</u>
force, 17, 20, **23**
  au of, <u>17</u>
  Lorentz, 20
  moment of, **23**
formal potential, 60
formation, 45, 47, 48, 50, **52**
formula, **46**, 47
  chemical, 45–47
formula entity, 48
formula unit, 47, 59
fraction
  amount, **44**, 50, 52, 53
  amount-of-substance, **44**
  mass, **44**
  mole, 4, **44**
  volume, **44**
fractional coordinate, 41
fractional population, 37

- franklin, 18
- free spectral range, 37
- frequency, **22**, 32, 36, 37, 57
    - angular, **22**, 36
    - collision, **55**
    - Debye, 41
    - Larmor, 27
    - standardized resonance, 32
    - transition, 29
- frequency factor, 55
- frequency unit, 29, **29**
    - angular, 31
- friction factor
    - dynamic, 23
- fugacity, **50**, 62
- fugacity coefficient, 50
- function
    - (atomic) basis, 26
    - configuration, 26
    - Gibbs, 49
    - Hamilton, 23
    - Heaviside, **9**, 43
    - Legendre, 25
    - spherical harmonic, 25
    - work, **41**
- fundamental physical constant, 6, 13
- fundamental translation vector, 41
    - angular, 41
- fundamental vibration, 29
- fundamental wavenumber
    - vibrational, 29
- fusion (subscript), 51
- FWHM, **37**, 57

- *g*-factor, 27, 31, **31**
    - Landé, 27, 31, **31**
    - nuclear, 27
- Galvani potential difference, **59**, 60
- galvanic cell, 61, **61**
- gamma function, 9
- gas, **37**, 39, 48, 62, 64
    - ideal, 39, **52**, 56
- gas constant, 43
    - molar, 43
- gas mixture, **52**
- gas phase, 48, 52
- gauss, 19, 19
- Gauss law, 20
- Gaussian, 18
- general chemical equation, 45
- general chemistry, 44–48
- general rules for symbol, 3
- general wavefunction, 26
- generalized coordinate, **22**, 23, 43
- generalized first-order kinetics, 56
- generalized momentum, 23, **43**
- Gibbs dividing surface, **63**
- Gibbs energy, 4, **49**, 61
    - partial molar, 59
    - standard, 59
    - standard reaction, 50
- Gibbs energy of activation
    - standard, 56
- Gibbs function, 49
- Gibbs surface, **63**
- gibi, 14
- giga, 14
- gigabinary, 14
- gradient
    - electric field, **31**
- gradient of a scalar field, 9
- gradient tensor, 31
- gram, 14
- Grashof number (mass transfer), 65
- gravitational constant, ifc
- gray, 17
- Greek alphabet, 76
- Greek letter, 6, 34, 48
- ground electronic state, 35
- ground state, **6**, 27, 35, 52
- ground-state distance, 30
- ground-state label, 35
- gyromagnetic ratio, **27**, 31, 32

- Hückel molecular orbital theory, 26
- half cell, **61**, 62
- half life, 28, 55
- Hamilton function, 23
- Hamiltonian, **25**
    - core, 26
    - hyperfine coupling, 31
    - rotational, 30
- Hamiltonian operator, *see* Hamiltonian
- hard sphere collision model, 55
- hard sphere radius, 55
- harmonic (vibrational) wavenumber, 29, **29**
- hartree, Hartree energy ifc, 6, **13**, 15, 17
- Hartree–Fock–Roothaan, 26
- heat, **49**, 61
- heat capacities
    - ratio of, 49
- heat capacity, 3, **49**
- heat capacity (constant pressure), 3, 4, **49**
    - specific, **4**, 65
- heat capacity (constant volume), **49**
- heat flux, 64
- heat flux density, **64**
- heat transfer
    - coefficient of, 64, 65
- Heaviside function, **9**, 43
- hecto, 14
- height, 22
- Helmholtz energy, **49**
- Henry's law, 50

Henry's law constant, 50, **50**
Hermitian conjugate, **25**
hermitian conjugate, 9
hertz, 4
high energy, 36
high pressure limit, 56
HMO, 26
horse power
  metric, 17
hour, 16
hydrogen-like atom, 27
hydrogen-like wavefunction, 25
hyperfine coupling constant, 31
hyperfine coupling Hamiltonian, 31

ideal (superscript), 51
ideal gas, 39, **52**, 56
identity operator, 33, 34
IEC, 4, **14**
imaginary refractive index, **38**
immersion (subscript), 51
inch, 16
incident intensity, 37
indirect mechanism, 57
induced emission, 37, **38**
inductance, 19, **24**
  self-, 19, **24**
induction
  magnetic, **24**, 31
inexact differential, 49
infinite dilution, **48**, 52
infinite dilution (superscript), 51
inner potential difference, 59
integral
  Coulomb, 26
  exchange, 26
  one-electron, 26
  two-electron, 26
integrated absorption cross section, 38
integrated net absorption cross section, 38
integration element, 25
intensity, 36, 37
  absorption, 39
  band, 37, 39
  incident, 37
  line, 37, 39
  luminous, 2, 11, **36**
  photon, 36
  radiant, **36**
  spectral, **36**, 37
  transmitted, 37
intensity of electromagnetic radiation, 36
intensive, **4**, 64
interatomic equilibrium distance, 30
interatomic (internuclear) distance, 30
interfacial tension, **63**
internal absorptance, 39
internal energy, **49**
internal vibrational angular momentum, 33
internal vibrational coordinate, 30
international calorie, 17
inverse of a square matrix, 9
inversion operation, **34**
inversion operator, 34
ion, 46
ion density, 41
ionic activity
  mean, 59
ionic activity coefficient
  mean, 59
ionic charge number, 46
ionic concentration
  mean, 59
ionic conductivity, 48, **60**
ionic conductor, 61
ionic molality
  mean, 59
ionic strength, **50**, 59
ionic strength (concentration basis), **50**, 59
ionic strength (molality basis), **50**, 59
ionization energy, 27
ionization potential, 27
irradiance, 36, **36**
  spectral, 36
ISO, 4, 21, **21**, 36, 49, 64
ISO 1000, 5
ISO/IEC 80000, 5
ISO/TC 12, viii
isobaric nuclide, 46
isobars, 46
isoelectric point, 63
isotope, **46**
isotopic nuclide, 46
ISQ, **2**
italic fonts, 5
IUPAC, **viii**, 2, 4, 21, 29, 31, 36, 53, 54, 63
IUPAP, **viii**, 21, 36, 49, 64

Josephson constant, ifc
joule, 4, 12
Joule–Thomson coefficient, 49

katal, 58
kelvin, 11, 17
kibi, 14
kilo, 14
kilobinary, 14
kilodalton, 12
kilogram, 11, 14, 16
kilogram-force, 17
kilometre, 14
kilopond, 17
kilovolt, 4
kilowatt-hour, 17

kinematic viscosity, **23**
kinetic energy, 12, **23**
kinetic energy operator, 25
kinetics, **54**
  chemical, **54**, 57
Kronecker delta, 8

label for symmetry species, 33
Lagrangian, 43
Landé *g*-factor, ifc, 27, 31, **31**
langmuir, 64
Laplacian operator, 9
Larmor angular frequency, 27
Larmor circular frequency, 27
laser cavity, 37
Latin letter, 34
lattice constant, 41
lattice direction, 42
lattice parameter, 41
lattice plane spacing, 41
lattice vector, 41
  angular reciprocal, 41
  Bravais, 41
  reciprocal, **41**
Legendre function, 25
length, 1, 2, 4, 11, 13, 16, **22**, 26, 41, 43, 47, 65
  absorbing path, 39
  characteristic Debye, 63
  Debye, 63
  Debye screening, 63
  equilibrium bond, 13
  path, 37, **38**
  reciprocal unit cell, 41
  unit cell, 41
length of arc, 22
level
  energy, 37
  macroscopic, **46**, 47
  microscopic, **45**, 46, 47
level of confidence, **68**
Levi-Civita symbol, 9
lifetime, 28, 55
  natural, 56
light, 36
  speed of, 6, 13, 22, 29, **36**, 41
light source, 38
light transmitted, 39
line intensity, 37, 39
line width, 37
line width at half maximum, 37
linear decadic absorption coefficient, 38
linear molecule, 33, 34
linear Napierian absorption coefficient, 38
linear strain, 23
linear top, 33
lineic, 4
lineshape, 37
linewidth, 57
  natural, 56
  predissociation, 56
liquid, 31, 38, 48, 61
liquid junction, 61
liquid junction potential, 60
liquid phase, 48
liquid state, 52
litre, 4, 16
logarithm
  decadic, 54
logical operator, 9
longitudinal cavity mode, 37
longitudinal spin-lattice relaxation time, 32
Lorentz force, 20
luminescence, **37**, 39
luminous efficacy, 11
luminous intensity, 2, 11, **36**
luminous quantity, 36

Mach number, 65
macroscopic level, **46**, 47
macroscopic magnetization
  equilibrium, 31
magnetic constant, ifc, **24**
magnetic dipole, 20
magnetic dipole moment, 13, 19, 24, **27**
  atomic unit of, 13
magnetic field, 32
magnetic field strength, 19, 20, **24**
magnetic flux, 19, **24**
magnetic flux density, 13, 19, 20, **24**, 31
  effective, 32
magnetic induction, **24**, 31
magnetic quantum number, 33
magnetic susceptibility, 19, 20, **24**
  molar, 19, **24**
magnetic vector potential, **24**
magnetism, 24
magnetizability, 19, **24**, **27**
magnetization, 20, **24**, 31
magnetization (volume), 19
magnetogyric ratio, *see* gyromagnetic ratio
magneton
  Bohr, 13, 19, **27**, 32
  nuclear, 19, **27**, 32
mass, 2, 4, 5, 11–13, 16, **23**, 26, 31, 43, 45, 63
  atomic, **27**, 44
  centre of, 34
  electron, 6, 12, **13**, 16, 32
  molar, **44**, 48
  proton, 32
  reduced, 6, **23**, 55
  relative atomic, **44**
  relative molar, 44
  relative molecular, 44
  SI base unit of, 14

- mass adjusted vibrational coordinate, 30
- mass concentration, 38, **44**
- mass constant
  - atomic, 44
  - molar, **44**
- mass density, 4, **23**, 38, 47
- mass fraction, **44**
- mass number, **27**, 45, 46
- mass of atom, **27**
- mass of entity, 44
- mass transfer
  - Nusselt number for, 65
  - steady-state, 60
- mass transfer coefficient, **60**, 65
- mass unit
  - unified atomic, ifc, 6, 12, 16, 44
- massic, 4
- massic volume, 4
- mathematical constant, 5, 7
- mathematical function, 7
- mathematical operator, 5, **8**
  - symbol of, 5
- mathematical symbol, 5, 7–9
  - printing of, **7**
- matrix, 3, 5, **9**
  - complex conjugate, 9
  - density, 43
  - determinant of a square, 9
  - inverse of square, 9
  - rate (coefficient), 56
  - scattering, 55
  - trace of a square, 9
  - transpose of, 9
  - unit, 9
  - unitary, 55
- matrix element, **9**, 25, 56
  - density, 26, 43, **43**
  - overlap, 26
- matrix element of operator, 25
- matrix equation, 56
- matrix notation, **35**
- maxwell, 19
- Maxwell equations, 20
- mean, 67
  - standard deviation of the, 67
- mean ionic activity, 59
- mean ionic activity coefficient, 59
- mean ionic concentration, 59
- mean ionic molality, 59
- mean life, 55
- mean relative speed, 55
- mean value, 8
- measure of uncertainty, 68
- measured quantity, 67
- mebi, 14
- mechanics
  - classical, 23
  - quantum, 25
  - statistical, 56
- mechanism
  - catalytic reaction, 57
  - chain-reaction, 57
  - complex, 57
  - composite, 57
  - indirect, 57
  - stepwise, 57
- mega, 14
- megabinary, 14
- megadalton, 12
- megahertz, 4
- melting (subscript), 51
- metallic conductor, 61
- metre, 1, 4, 11, 16, 47
  - cubic, 16
  - square, 16
- metric horse power, 17
- Michaelis constant, 56, **56**
- micro, 14
- microcanonical ensemble, 43
- microgram, 6
- micromolar, 45
- micron, 16
- microscopic level, **45**, 46, 47
- migration velocity, 60
- mile, 16
- Miller indices, **42**
- milli, 14
- millimetre of mercury, 17
- millimolar, 45
- minute, 16
- mixing of fluids (subscript), 51
- mixture, 43, **50**
  - gas, **52**
- mobility
  - electric, 60
- modulus
  - bulk, 23
  - compression, 23
  - Coulomb's, 23
  - shear, 23
  - Young's, 23
- modulus of elasticity, 23
- modulus of the Poynting vector, 36
- molality, **45**, 50, 52, 53, 59
  - mean ionic, 59
  - standard, 52, **53**
- molar, **4**, 45
  - partial, 49
- molar absorption coefficient, 4
- molar conductivity, 4, 48, **60**
- molar conductivity of an ion, 60
- molar decadic absorption coefficient, 38, 39

molar enthalpy, **4**
  standard partial, 50
molar enthalpy of formation
  standard, 50
molar enthalpy of reaction, 51
molar enthalpy of vaporization, 51
molar entropy, **48**
molar entropy of reaction, 51
molar excitation energy, 15
molar gas constant, ifc, 43
molar Gibbs energy
  partial, 4, 59
molar magnetic susceptibility, 19, **24**
molar mass, **44**, 48
  relative, 44
molar mass constant, ifc, **44**
molar Napierian absorption coefficient, 38
molar optical rotatory power, 38, **38**
molar quantity, 4, **49**
  partial, 4, **49**
molar susceptibility
  non-rationalized, 19
molar volume, 4, **44**, 48
  partial, 49
molarity, **45**
mole, 2, 11, 47
mole fraction, 4, **44**
molecular mass
  relative, 44
molecular momentum vector, 43
molecular orbital, 26
molecular partition function
  (per volume), 56
  (transition state), 56
molecular point group, 34
molecular position vector, 43
molecular quadrupole moment, 27
molecular quantity, 37
molecular spectroscopy, 23
molecular spin orbital, 26
molecular state, 34
molecular velocity vector, 43
molecular weight, 44
molecularity, 57, 58
molecule-fixed coordinate, 34
moment of force, **23**
moment of inertia, 23
  principal, 29
momentum, **23**, 43
  angular, 13, 17, **23**, 26, 31, 33
  generalized, 23, **43**
  spin angular, **25**, 31
momentum operator, 25
momentum transfer, **41**
momentum vector
  molecular, 43
monomolecular reaction, 57
muon, 6

nabla operator, **9**, 25
name
  chemical-compound, 6
name of a physical quantity, 1, **3**, 47
nano, 14
nanometre, **1**, 4
Napierian absorbance, **38**
Napierian absorption coefficient, 38, **38**
natural lifetime, 56
natural linewidth, 56
Nernst equation, **60**
net absorption coefficient, 37
net absorption cross section, 37
net forward reaction, 47
net integrated absorption band intensity, 39
net reaction, 57
neutron, 6, 45
neutron diffraction, 41
neutron mas, ifc
neutron number, **27**
newton, 17
Newtonian constant of gravitation, ifc
NMR, 32
nomenclature, 6
  biochemical, 6
  inorganic, 6
  macromolecular, 6
  organic, 6
non-rationalized molar susceptibility, 19
non-SI unit, 15, **16**
normal distribution, 68
normal mode, 35
normal stress, 23
normalization factor, **25**
normalized wavefunction, 25
nuclear chemistry, 36
nuclear g-factor, 27
nuclear magneton, ifc, 19, **27**, 32
nuclear orbital angular momentum (rotational), 33
nuclear quadrupole coupling constant, 32
nuclear quadrupole interaction energy tensor, 28
nuclear quadrupole moment, **27**, 31
nuclear spin angular momentum, 33
nuclear spin-spin coupling, 31
nuclear spin-spin coupling constant
  reduced, 31
nucleon number, **27**, 46
nuclide, **46**
  isobaric, 46
  isotopic, 46
number, **6**, 22, 28, 54
  atomic, 27, **27**, 46
  Avogadro, 44
  characteristic, 3

  charge, 26, 46, 59
  collision, 55
  electron, 59
  Grashof, 65
  ionic charge, 46
  Mach, 65
  mass, **27**, 45, 46
  neutron, **27**
  nucleon, **27**, 46
  oxidation, 46
  Péclet, 65
  Prandtl, 65
  proton, 26, **27**, 46
  quantum, 7, 31, **33**, 34, 35, 47
  Reynolds, 3, **65**
  Schmidt, 65
  Sherwood, 65
  Stanton, 65
  stoichiometric, 45, **45**, 47, 51, 52, 59, 60
  symmetry, 43
  transport (characteristic), 65
  Weber, 65
number concentration, 44, 54, 55
number flux density, 64
number of adsorbed molecules, 63
number of anions, 59
number of atoms, 41, 46
number of cations, 59
number of collisions
  total, 55
number of defined events, 57
number of electrons, **59**, 60
number of entities, 4, 44
number of measurements, 67
number of moles, 47
number of particles, 57
number of photons absorbed, 57
number of quantum states, 43
number of states, **56**
  cumulative, 43
numbers
  printing of, 7
numerical value of a physical quantity, **1**
Nusselt number (mass transfer), 65
nutation angular frequency, 31

oersted, 19
ohm, 18
one-electron integral, 26
one-electron orbital, 34, 35
operator, **7**, 25
  angular momentum, 25, **33**
  del, 9
  density, 43
  electron orbital, 30
  electron spin, 30, **31**
  expectation value of, 25
  Fock, 26
  Hamiltonian, 25
  Hermitian conjugate of, 25
  identity, 33, 34
  inversion, 34
  kinetic energy, 25
  Laplacian, 9
  logical, 9
  mathematical, 5, **8**
  matrix element of, 25
  momentum, 25
  nabla, **9**, 25
  parity, 34
  permutation, 33
  permutation-inversion, 33
  reflection, 34
  rotation, 34
  rotation-reflection, 34
  space-fixed inversion, 33
  symmetry, 33, 34
  vector, 33
operator symbol, 33
operator symbol (reaction in general), 50
optical rotation, 38
  angle of, 38, **38**
optical rotatory power
  molar, 38, **38**
  specific, 38, **38**
optical spectroscopy, 34, 37
orbital, 26
  molecular, 26
  molecular spin, 26
  spin, 26
orbital angular momentum, 34
orbital energy, 26
order (Legendre function), 25
order of reaction, **56**, 60
  overall, 54
  partial, **54**
oscillator strength, 38
overall order of reaction, 54
overlap matrix element, 26
overpotential, 60
oxidation number, 46
oxidation rate constant, 60
oxidation state, 46

parameter
  energy, 26
parity, **34**
parity operator, 34
partial molar, 49
partial molar Gibbs energy, 4, 59
partial molar quantity, 4, **49**
partial molar volume, 49
partial order of reaction, **54**
partial pressure, 3, **44**, 54

particle, 41, 48
α-, 6
particle symbol, 6
partition function, 37, **43**, 56
partition function (per volume), 56
partition function (transition state)
molecular, 56
pascal, 17
path, **22**
path length, 37, **38**
pebi, 14
period, 22
periodic, 22
permeability, 6, **24**
relative, 20, **24**
permittivity, **24**
relative, 20
permutation operator, 33
permutation-inversion operator, 33
peta, 14
petabinary, 14
pH, 3, 59
definition of, 59
phase
condensed, 49
fluid, 48
phase space
volume in, 43
photochemical yield, 56
photochemistry, 36, 54–57
photon, 57
photon absorption
rate of, 57
photon intensity, 36
photon quantity, 36
physical constant
fundamental, 6, 13
physical quantity, **1**, 2, 6, 12, 15, 16
extensive, 64
intensive, 64
name of, 1, **3**, 47
numerical value of, **1**
product of, 5
quotient of, 5
symbol of, **3**, 5, 6, 15, 21
table of, 21
unit of, **1**, 25
value of, **1**, 15
pico, 14
Planck constant, ifc, 6, 11, 13
plane angle, 16, **22**, 38
plane of polarization, 38
plane of rotation, 22
point charge, 24
point group, 7
point of zero charge, 63
poise, 17
polar coordinate
spherical, **22**
polarizability, 18, **28**
electric, 13
polarizability of a molecule
electric, 28
polarizability volume, 18
polarization, 19
dielectric, 20, **24**
volume, 19
polytropy, 43
position vector, 22, 41
molecular, 43
potential
cell, **59**
chemical, 3, 4, **49**, 52
electric, 18, **24**
electrochemical, 59
electrode, **59**
electrokinetic, 60
electromagnetic, **20**
equilibrium, **59**
formal, 60
ionization, 27
liquid junction, 60
standard, **62**
standard cell, 62
standard chemical, **49**, 52, 53
ζ-, 60
potential (of cell reaction)
standard, **59**, 61, 62
potential barrier, 12
potential difference
electric, **24**, 61
Galvani, **59**, 60
inner, 59
potential difference (electrochemical cell), 61
potential energy, **23**
potential of electrochemical cell reaction, 61
power, 17, **23**
radiant, 36
resolving, 37
thermal, **49**, 64
power per area
radiation, 36
Poynting vector, 20, **24**, 36
Prandtl number, 65
pre-exponential factor, 54, **55**
predissociation linewidth, 56
predissociation rate constant, 57
prefix (symbol), 14
pressure, 4, 17, **23**, 39, 44, 64
atmospheric, 49
partial, 3, **44**, 54
standard, 52, **53**, 62

standard state, 53
surface, 63
total, 44
vapor, 15, **51**
pressure change
rate of, 54
pressure dependence (of the activity), 49
pressure virial coefficient, 49
primary symmetry axis, 34
principal component, 32
principal moment of inertia, 29
principal quantum number, 27
printing of mathematical symbols, **7**
printing of numbers, 7
probability, 43, **43**, 55
Einstein transition, 39
estimated, 68
transition, 55
probability current density, 25
probability density, **25**
process, 51
transport, 54
process symbol, 51
product of unit, 5
proton, 6, 32, 45
proton charge (au), 18
proton mass, ifc, 32
proton number, 26, **27**, 46
pure phase, 38, 50, **52**
pure state, 43
pure substance, 49, **52**
pure substance (superscript), 51

quadrupole coupling constant, 31, 32
quadrupole interaction energy tensor, 28
nuclear, 28
quadrupole moment
electric, **13**, 18
molecular, 27
nuclear, **27**, 31
quadrupole moment (of a molecule), 27
quadrupole moment (of a nucleus), 27
quantity
base, **2**, 11
derived, 2
dimensionless, 3, 21
extensive, **4**, 49, 64
extensive thermodynamic, 51
luminous, 36
molar, 4, **49**
partial molar, 4, **49**
photon, 36
physical, *see* physical quantity
specific, **4**, 49
tensor, 23
quantity calculus, 1, 15
quantum chemistry, 13, 25
quantum mechanics, 25
quantum number, 7, 31, **33**, 34, 35, 47
angular momentum, **33**
azimuthal, 33
component, 33
magnetic, 33
principal, 27
vibrational, **29**, 35
quantum yield, **56**, 57
quotient of unit, 5

radian, 16, 22
radiance, 36
radiant energy, 36
radiant energy density, 36
spectral, 36, **36**
radiant intensity, **36**
radiant power, 36, 37
radiation energy flow
rate of, 20
radiation power per area, 36
radical, 46
radioactive activity, 17
radius, **22**
Bohr, ifc, 6, 13
hard sphere, 55
Rankine temperature, 17
Raoult's law, 50
rate, 64
reaction, 54
rate (coefficient) matrix, 56
rate coefficient, *see* rate constant
rate constant, 54, 55, 58
first-order, 56, **56**
fluorescence, **56**, 57
oxidation, 60
predissociation, 57
reduction, 60
rate constant of fluorescence, 57
rate constant of unimolecular reaction (high pressure), 54
rate constant of unimolecular reaction (low pressure), 54
rate equation, 56
rate law, 54, 58, **58**
rate of change of quantity, 54
rate of concentration change, 54
rate of conversion, **54**, 57
rate of effusion, 64
rate of entropy change, 54
rate of photon absorption, 57
rate of pressure change, 54
rate of radiation energy flow, 20
rate of reaction (amount concentration), 54
rate of reaction (number concentration), 54
ratio of heat capacities, 49
reactance, 24

reactants
  zero-point level, 56
reaction, 47
  bimolecular, **57**, 58
  electrochemical, 59–61, **62**
  monomolecular, 57
  symbol for, 50
  termolecular, 57
  trimolecular, **57**, 58
  unimolecular, **57**, 58
reaction enthalpy, 58
  standard, 3, **50**
reaction entropy
  standard, 50
reaction equation, **47**, 52, 54, 58
reaction in general (subscript), **50**, 51
reaction order, **54**, 56, 60
reaction quantity
  standard, 51
reaction quotient, **50**, 52
reaction rate, 54, 58, **58**
reciprocal energy parameter, 43
reciprocal lattice vector, **41**
reciprocal unit cell angle, 41
reciprocal unit cell length, 41
reciprocal wavelength, **29**
reduced coupling constant, 31
reduced limiting sedimentation coefficient, 63
reduced mass, 6, **23**, 55
reduced nuclear spin-spin coupling constant, 31
reduction rate constant, 60
reference state, 52
reference state (of an element), 53
reflectance, 37
reflection, **37**, 39
reflection factor, 37
reflection loss, 37
reflection operator, 34
refraction index, 36
refractive index, **36**
  complex, **38**
  imaginary, **38**
relative activity, **49**
relative atomic mass, **44**
relative density, 23
relative molar mass, 44
relative molecular mass, 44
relative permeability, 20, **24**
relative permittivity, 20
relative speed
  mean, 55
relative standard uncertainty, 67
relaxation, 31
relaxation time, 22, 32, **55**
  longitudinal spin-lattice, 32
  spin-lattice, 32
repetency, 15, 29
resistance
  electric, 18, **24**
  thermal, 64
resolution, 37
resolving power, 37
resonance, 32
  electron paramagnetic, 31
  electron spin, 31
resonance frequency, 32
  standardized, 32
resonance structure, 47
Reynolds number, 3, **65**
roman fonts, 5
rotation
  specific, 38
rotation of a vector field, 9
rotation operator, 34
rotation-reflection operator, 34
rotational band, 35
rotational constant, 29
rotational constant (in frequency), 29
rotational constant (in wavenumber), 29
rotational Hamiltonian, 30
rotational state, **33**, 47
rotational transition, 35
rotatory power
  molar optical, 38, **38**
  specific optical, 38, **38**
rydberg, Rydberg constant, ifc, 17

scalar product, 9
scattering, **37**, 39
scattering factor
  atomic, 41
scattering matrix, 55
Schmidt number, 65
Schönflies symbol, 34
second, 4, 11, 16
second virial coefficient, 49
second radiation constant, ifc
second-rank tensor, **3**
sedimentation
  speed of, 63
sedimentation coefficient, 63, **63**
  reduced limiting, 63
self-inductance, 19, **24**
semiconductor space-charge theory, 63
SHE, 59, **62**
shear modulus, 23
shear strain, 23
shear stress, 23
shear stress tensor, 64
shear viscosity, **64**
Sherwood number, 65
shielding constant, 32
shielding tensor, 32

SI, 2, 11
SI base unit, 2, 11, 14
 symbol for, 11
SI derived unit, 14
SI prefix, 12, **14**
SI unit, 13
sievert, 17
solid, 31, 38, **48**
solid adsorbent, 63
solid angle, **22**, 36
solid phase, 48, 50, **52**
solid state, 41–52
solubility, **45**
solubility product, 50
solute, 45, 50, 52, **52**, 53
solution, 38, 48, 49, **52**, 60, 62
 aqueous, 48, **48**, 49, 62
solution (subscript), 51
solution at infinite dilution
 aquaeous, **48**
solvent, 39, 45, 50, **52**
sound
 speed of, **22**, 65
space and time, 22
space charge, 63
space coordinate
 Cartesian, **22**
space-fixed component, **33**
space-fixed coordinate, 33
space-fixed inversion operator, 33
space-fixed origin, 34
specific, 4
specific heat capacity (constant pressure), **4**, 65
specific optical rotatory power, 38, **38**
specific quantity, **4**, 49
specific rotation, 38
specific surface area, 63
specific volume, **4**, 23
spectral intensity, **36**, 37
spectral irradiance, 36
spectral radiant energy density, 36, **36**
spectroscopic transition, 35
spectroscopy, 29–35, 37
 atomic, 23
 molecular, 23
 optical, 34, 37
spectrum
 absorption, 37
speed, 12, 13, **17**, 22, 65
 average, **43**
 mean relative, 55
speed, **17**
speed distribution function, 43
speed of light, 6, 13, 22, 29, **36**, 41
speed of light in vacuum, ifc, 11
speed of sedimentation, 63
speed of sound, **22**, 65
spherical harmonic function, 25
spherical polar coordinate, **22**
spherical top, 33
spin angular momentum, **25**, 31
spin-lattice relaxation time, 32
 longitudinal, 32
spin multiplicity, 34
 electron, 34
spin operator
 electron, 30, **31**
spin-orbit coupling constant, 30
spin orbital, 26
 molecular, 26
spin-rotation coupling constant, 31
spin-spin coupling constant
 nuclear, 31
 reduced nuclear, 31
spin statistical weight, 43
spin wavefunction, **25**
 electron, 25
spontaneous emission, **38**, 57
square metre, 16
standard acceleration of gravity, ifc, 17
standard atmosphere, 17
standard cell potential, 62
standard chemical potential, **49**, 52, 53
standard concentration, **53**
standard condition, **52**, 59
standard deviation of the mean, 67
standard electrode potential, **59**, 62
standard enthalpy of activation, 55
standard enthalpy of combustion, 52
standard enthalpy of formation, 48
standard equilibrium constant, 52
standard Gibbs energy, 59
standard Gibbs energy of activation, 56
standard hydrogen electrode (SHE), 59, **62**
standard molality, 52, **53**
standard molar enthalpy of formation, 50
standard partial molar enthalpy, 50
standard potential, **62**
standard potential (of cell reaction), **59**, 61, 62
standard pressure, 52, **53**, 62
standard reaction enthalpy, 3, **50**
standard reaction entropy, 50
standard reaction Gibbs energy, 50
standard reaction quantity, 51
standard state, 52, **52**, 59
 biochemical, 53
 choice of, 50
standard state (gas phase), 52
standard state (liquid or solid state), 52
standard state pressure, 53
standard state (solute), 52
standard (symbol), 49, **51**, 59

standard uncertainty, 7, **67**
combined, 67, 68
relative, 67
standardized resonance frequency, 32
state
configuration, 26
state function, *see* wavefunction
state of aggregation, **48**
statistical entropy, 43
statistical mechanics, 56
statistical thermodynamics, 43
statistical weight, 29, **43**
spin, 43
steady-state mass transfer, 60
Stefan–Boltzmann constant, ifc, **37**
stepwise mechanism, 57
steradian, 22
stimulated emission, 38
Stockholm convention, 61
stoichiometric chemical equation, **51**, 58
stoichiometric equation, 45, **47**, 51, 54, 58
stoichiometric number, 45, **45**, 47, 51, 52, 59, 60
stoichiometric reaction equation, 51
stoichiometry, 58
strain
bulk, 23
linear, 23
shear, 23
volume, 23
stress
normal, 23
shear, 23
stress tensor
shear, 64
structure factor, 41
sublimation (subscript), 51
substance concentration, 2
sum of states, 56
sum over states, *see* partition function
surface area, **63**
specific, 63
surface charge density, 4
surface chemistry, 63
surface concentration, 54, **63**
surface coverage, **63**
surface density, 23
surface excess, **63**
surface excess amount, **63**
surface excess concentration, 63
surface layer, 63
surface pressure, 63
surface tension, **23**, 63, 65
susceptibility
electric, 20, **24**
magnetic, 19, 20, **24**
molar magnetic, 19, **24**
symbol
fonts for, 5
general rules for, 3
mathematical, 5, 7–9
Schönflies, 34
symbol of
angular momentum operator, 33
atomic unit, 13
base quantity, 2
chemical element, 6, **45**, 46
chemical thermodynamics, 49–53
decimal (sub)multiple, 14
dimension, **2**
directions in crystal, 42
excitation, 46
mathematical operator, 5
particle, 6
physical or chemical process, 51
physical quantity, **3**, 5, 6, 15, 21
planes in crystals, 42
prefix, 14
reaction, 50, **51**
SI base unit, 11
SI derived unit, 14
standard state, 49, **51**, 59
symmetry operator, 33
symmetry species, 7
term, 34, 35
unit, **1**, 2–4, 11
symmetric reduction, 30
symmetric top, **33**, 35
oblate, 33
prolate, 33
symmetry, 6
symmetry axis, 33
symmetry number, 43
symmetry operator, 33, 34
symmetry species, 34, 35
label for, 33
symbol for, 7

Tafel slope, 60
tebi, 14
temperature, 4, 37, 39, 52, 53
Celsius, 17, **49**
centigrade, 49
Debye, **41**
Fahrenheit, 17
Rankine, 17
thermodynamic, 2, 11, 17, 65
tension
electric, **24**
interfacial, **63**
surface, **23**, 63, 65
tensor, **3**, 5, 6, 27, 31
electric field gradient, 28
field gradient, 32

  gradient, 31
  quadrupole interaction energy, 28
  second-rank, **3**
  shear stress, 64
  shielding, 32
tensor quantity, 23
tera, 14
terabinary, 14
term, 36
  total, 29
term formula
  vibrational, 29
term symbol, **35**
  electronic, 46
term symbol (atomic state), 34
term symbol (molecular state), 34
term value, 29
termolecular reaction, 57
tesla, 19
thermal averaging, 41
thermal conductance, 64
thermal conductivity, **64**
thermal diffusion coefficient, 64
thermal diffusivity, 64, 65
thermal equilibrium, 37
thermal power, **49**, 64
thermal resistance, 64
thermochemical calorie, 17
thermodynamic equilibrium constant, 50
thermodynamic temperature, 2, 11, 17, 65
thermodynamics
  chemical, 49–53
  statistical, 43
thickness, **22**, 63
  diffusion layer, 60
third virial coefficient, 49
three-body collision, 55
threshold energy, 56
time, 2, 11, 13, 16, **22**, 64
  au of, 16
  correlation, 32
  decay, 55
  relaxation, 22, 32, **55**
time integral of fluence rate, 37
time interval, 37, **55**
top
  asymmetric, **33**, 35
  linear, 33
  spherical, 33
  symmetric, **33**, 35
torque, 23
Torr, 15, 17, 64
torr, 15, 17
total angular momentum, 34
total pressure, 44
total term, 29
trace of a square matrix, 9
transfer coefficient
  mass, **60**, 65
transition
  electronic, 38
  rotational, 35
  spectroscopic, 35
  vibrational, 35
  vibronic, 35
transition between phases (subscript), 51
transition dipole moment (molecule), 30
transition frequency, 29
transition probability, 55
  Einstein, 39
transition state
  (molecular) partition function, 56
  partition function, 56
  zero-point level, 56
transition state (superscript), 51
transition state theory, 55, 56
transition wavelength, 38
transition wavenumber, 29
translation vector
  (angular) fundamental, 41
transmission coefficient, 56
transmission factor, 37
transmittance, 37
transmitted intensity, 37
transport (characteristic) number, 65
transport process, 54
transport properties, 64–65
transpose of matrix, 9
trimolecular reaction, **57**, 58
triple point (subscript), 51
two-electron integral, 26
type A evaluation, 67
type B evaluation, **67**

uncertainty, 7, 67–68
  expanded, 67, 68
  standard, 7, **67**
unified atomic mass unit, 6, 12, 16, 44
unimolecular reaction, **57**, 58
unit
  angular frequency, 31
  atomic, 2, 13, 15, 16, 26
  base, 11–36
  cgs, 13, 16–17
  choice of, 1
  coherent, 12
  decimal multiple of, 4, **14**
  decimal submultiple of, 4, **14**
  frequency, 29, **29**
  product of, 5
  quotient of, 5
  SI base, 2, 11, 14
  SI derived, 14

symbol for, **1**, 2–4, 11
unit cell angle, 41
unit cell length, 41
unit matrix, 9
unit of physical quantity, **1**, 25
unit step function, 9
unit vector, 22
unitary matrix, 55
value of physical quantity, **1**
van der Waals coefficient, 49
vapor, 48
vapor pressure, 15, **51**
vaporization (subscript), 51
vector, 3, 5, **9**, 22, 24
dipole, 24
fundamental translation, 41
lattice, 41
position, 22
Poynting, 20, **24**, 36
unit, 22
vector operator, 33
vector potential
magnetic, **24**
vector product, 9
velocity, 20, **22**
angular, **22**
migration, 60
velocity distribution function, 43
velocity vector
molecular, 43
vibration
fundamental, 29
vibrational angular momentum
internal, 33
vibrational anharmonicity constant, 29
vibrational coordinate
dimensionless, 30
internal, 30
mass adjusted, 30
vibrational displacement, 30
vibrational fundamental wavenumber, 29
vibrational ground state, 29
vibrational modes, 29
vibrational normal coordinate, 30
vibrational quantum number, **29**, 35
vibrational state, **35**, 47
vibrational term formula, 29
vibrational transition, 35
vibronic transition, 35
virial coefficient, 49
viscosity, 64, 65
bulk, 64
dynamic, 17, **23**
kinematic, **23**
shear, **64**
volt, 18
volume, 4, 16, **22**, 54, 64
massic, 4
molar, 4, **44**, 48
partial molar, 49
polarizability, 18
specific, **4**, 23
volume fraction, **44**
volume in phase space, 43
volume magnetization, 19
volume of activation, 55
volume polarization, 19
volume strain, 23
volumic, 4
von Klitzing constant, ifc

watt, 17
wavefunction, 25, 26
general, 26
hydrogen-like, 25
normalized, 25
spin, **25**
wavelength, 1, 36
reciprocal, **29**
transition, 38
wavenumber, 15, **29**, 36–39
Debye, 41
harmonic (vibrational), 29, **29**
transition, 29
vibrational fundamental, 29
wavenumber (in vacuum), 36
weber, 19
Weber number, 65
weight, 23
atomic, **44**
molecular, 44
spin statistical, 43
statistical, 29, **43**
work, 23, **49**
work function, **41**
electron, 41

yard, 16
year, 16
yield
photochemical, 56
quantum, **56**, 57
yobi, 14
yocto, 14
yotta, 14
yottabinary, 14
Young's modulus, 23

zebi, 14
zepto, 14
zero of Celsius scale, ifc
zero-point level (reactants), 56
zero-point level (transition state), 56
ζ-potential, 60
zetta, 14
zettabinary, 14

PRESSURE CONVERSION FACTORS

	Pa	kPa	bar	atm	Torr	psi
1 Pa	$= 1$	$= 10^{-3}$	$= 10^{-5}$	$\approx 9.869\ 23\times10^{-6}$	$\approx 7.500\ 62\times10^{-3}$	$\approx 1.450\ 38\times10^{-4}$
1 kPa	$= 10^{3}$	$= 1$	$= 10^{-2}$	$\approx 9.869\ 23\times10^{-3}$	$\approx 7.500\ 62$	$\approx 0.145\ 038$
1 bar	$= 10^{5}$	$= 10^{2}$	$= 1$	$\approx 0.986\ 923$	≈ 750.062	≈ 14.5038
1 atm	$= 101\ 325$	$= 101.325$	$= 1.013\ 25$	$= 1$	$= 760$	≈ 14.6959
1 Torr	≈ 133.322	$\approx 0.133\ 322$	$\approx 1.333\ 22\times10^{-3}$	$\approx 1.315\ 79\times10^{-3}$	$= 1$	$\approx 1.933\ 68\times10^{-2}$
1 psi	≈ 6894.76	$\approx 6.894\ 76$	$\approx 6.894\ 76\times10^{-2}$	$\approx 6.804\ 60\times10^{-2}$	$\approx 51.714\ 94$	1

Examples of the use of this table: 1 bar $\approx$ 0.986 923 atm
1 Torr $\approx$ 133.322 Pa

Note: 1 mmHg = 1 Torr, to better than 2×10^{-7} Torr, see Section 3.2, p. 17.

NUMERICAL ENERGY CONVERSION FACTORS

$E = h\nu = hc\tilde{\nu} = kT$; $E_m = N_A E$

			wavenumber $\tilde{\nu}$	frequency ν	energy E			molar energy E_m		temperature T
			cm^{-1}	MHz	aJ	eV	E_h	$kJ\ mol^{-1}$	$kcal\ mol^{-1}$	K
$\tilde{\nu}$:	$1\ cm^{-1}$	$\widehat{=}$	1	$2.997\ 925\times10^{4}$	$1.986\ 446\times10^{-5}$	$1.239\ 842\times10^{-4}$	$4.556\ 335\times10^{-6}$	$11.962\ 66\times10^{-3}$	$2.859\ 144\times10^{-3}$	1.438 777
ν:	1 MHz	$\widehat{=}$	$3.335\ 641\times10^{-5}$	1	$6.626\ 070\times10^{-10}$	$4.135\ 668\times10^{-9}$	$1.519\ 830\times10^{-10}$	$3.990\ 313\times10^{-7}$	$9.537\ 076\times10^{-8}$	$4.799\ 243\times10^{-5}$
	1 aJ	$\widehat{=}$	50 341.17	$1.509\ 190\times10^{9}$	1	6.241 509	0.229 371 2	602.2141	143.9326	$7.242\ 971\times10^{4}$
E:	1 eV	$\widehat{=}$	8065.544	$2.417\ 989\times10^{8}$	0.160 217 7	1	$3.674\ 932\times10^{-2}$	96.485 33	23.060 55	$1.160\ 452\times10^{4}$
	1 E_h	$\widehat{=}$	219 474.6	$6.579\ 684\times10^{9}$	4.359 744	27.211 39	1	2625.500	627.5095	$3.157\ 750\times10^{5}$
E_m:	$1\ kJ\ mol^{-1}$	$\widehat{=}$	83.593 47	$2.506\ 069\times10^{6}$	$1.660\ 539\times10^{-3}$	$1.036\ 427\times10^{-2}$	$3.808\ 799\times10^{-4}$	1	0.239 005 7	120.2724
	$1\ kcal\ mol^{-1}$	$\widehat{=}$	349.7551	$1.048\ 539\times10^{7}$	$6.947\ 695\times10^{-3}$	$4.336\ 410\times10^{-2}$	$1.593\ 601\times10^{-3}$	4.184	1	503.2195
T:	1 K	$\widehat{=}$	0.695 034 8	$2.083\ 662\times10^{4}$	$1.380\ 650\times10^{-5}$	$8.617\ 333\times10^{-5}$	$3.166\ 812\times10^{-6}$	$8.314\ 463\times10^{-3}$	$1.987\ 204\times10^{-3}$	1

The symbol $\widehat{=}$ should be read as meaning 'approximately corresponding to' or 'is approximately equivalent to'. The conversion from kJ to kcal is exact by definition of the thermochemical kcal. The values in this table have been obtained from the constants on the inside front cover page. The last digit is given but may not be significant.

Examples of the use of this table:

$1\ aJ \ \widehat{=}\ 50\ 341.17\ cm^{-1}$

$1\ eV \ \widehat{=}\ 96.485\ 33\ kJ\ mol^{-1}$

Examples of the derivation of the conversion factors:

1 aJ to MHz

$$\frac{(1\ aJ)}{h} \ \widehat{=}\ \frac{10^{-18}\ J}{6.626\ 070\ 15\times10^{-34}\ J\ s} \ \widehat{=}\ 1.509\ 190\times10^{15}\ s^{-1} \ \widehat{=}\ 1.509\ 190\times10^{9}\ MHz$$

$1\ cm^{-1}$ to eV

$$(1\ cm^{-1})\ hc\left(\frac{e}{e}\right) \ \widehat{=}\ \frac{(1.986\ 445\ 857\times10^{-25}\ J)\times10^{2}\ e}{1.602\ 176\ 634\times10^{-19}\ C} \ \widehat{=}\ 1.239\ 842\times10^{-4}\ eV$$

$1\ E_h$ to $kJ\ mol^{-1}$

$$(1\ E_h)\ N_A \ \widehat{=}\ (4.359\ 744\ 72\times10^{-18}\ J)\times(6.022\ 140\ 76\times10^{23}\ mol^{-1}) \ \widehat{=}\ 2625.500\ kJ\ mol^{-1}$$

$1\ kcal\ mol^{-1}$ to cm^{-1}

$$\frac{(1\ kcal\ mol^{-1})}{hcN_A} \ \widehat{=}\ \frac{4.184\times(1\ kJ\ mol^{-1})}{hcN_A} \ \widehat{=}\ \frac{4.184\times(10^{3}\ J\ mol^{-1})}{(1.986\ 445\ 857\times10^{-25}\ J)\times10^{2}\ cm\times(6.022\ 140\ 76\times10^{23}\ mol^{-1})} \ \widehat{=}\ 349.7551\ cm^{-1}$$